GEL ELECTROPHORESIS: PROTEINS

ESSENTIAL TECHNIQUES SERIES

Series Editor
D. Rickwood
Department of Biological and Chemical Sciences, University of Essex, Wivenhoe Park, Colchester, UK

Published titles
Antibody Applications
Gel Electrophoresis: Nucleic Acids
DNA Isolation and Sequencing
Gel Electrophoresis: Proteins

Forthcoming titles
PCR
Gene Transcription
Human Chromosome Preparation
Cell Biology
Vectors: Cloning Applications
Vectors: Expression Systems
Cell Culture
Antibody Production

GEL ELECTROPHORESIS: PROTEINS

ESSENTIAL TECHNIQUES

D. M. Gersten

Guest Researcher, National Cancer Institute, National Institutes of Health, Bethesda, Maryland, USA

JOHN WILEY & SONS
Chichester • New York • Brisbane • Toronto • Singapore

Published in association with BIOS Scientific Publishers Limited

British Library Cataloguing in Publication Data
A catalogue record for this book is available from the British Library.

ISBN 0 471 962651

Library of Congress Cataloging in Publication Data
Gersten, Douglas
Gel electrophoresis of proteins : essential techniques / Douglas. Gersten.
p. cm. — — (Essential techniques series)
ISBN 0—471—96265—1 (alk. paper)
1. Proteins—Purification—Laboratory manuals. 2. Gel electrophoresis—Laboratory manuals. I. Title. II. Series.
QP551.G347 1996
574.19′245—dc20 95–49522
CIP

Typeset by Footnote Graphics Ltd, Warminster, UK
Printed and bound in UK by Biddles Ltd, Guildford, UK

CONTENTS

ABBREVIATIONS

BCA	bicinchoninic acid
bisacrylamide	*N,N'*-methylene bisacrylamide
BSA	bovine serum albumin
CCD	charge-coupled device
CHAPS	3-[3-(chloramidopropyl) dimethylammonio]-1-propanesulfonate
CHES	cyclohexylaminoethane sulfonic acid
c.p.m.	counts per minute
DAB	diamino benzidine
DBM	diazobenzomethyl
DMSO	dimethyl sulfoxide
DNase	deoxyribonuclease
d.p.m.	disintegrations per minute
DTT	dithiothreitol
EDTA	ethylenediamine tetraacetic acid
EEO	electroendosmosis
IEF	isoelectric focusing
IgG	immunoglobulin G
MTT	3-(4,5 dimethylthiazol-2-yl) 2,5-diphenyltetrazolium bromide
M_r	relative molecular weight
PMS	phenazine methosulfate
PMSF	phenylmethylsulfonyl fluoride
PPO	2,5-diphenyloxazole
PVDF	polyvinylidene difluoride
PVP	polyvinylpyrrolidone
RTV	room temperature vulcanizing
SDS	sodium dodecyl sulfate (sodium lauryl sulfate)
TCA	trichloroacetic acid
TEMED	*N,N,N',N'*-tetramethylethylene diamine
X-gal	5-bromo, 4-chloro, 3-indoyl-β-D-galactopyranoside

PREFACE

The aim of this volume is to provide clear, detailed descriptions of the most widely used protein electrophoresis techniques using numbered, step-by-step instructions. Explanatory notes on the opposite side of the page accompany each protocol. These consist of helpful hints, troubleshooting advice, safety information and points where the procedure may be interrupted.

The book is aimed primarily at predoctoral and postdoctoral researchers but it is also of value to more experienced workers, providing a convenient summary of essential techniques. Three points are taken into consideration in making the book appealing to the widest audience. Firstly, to serve 'first-time' users, particular attention is paid to initial aspects of protein electrophoresis such as gel casting. Secondly, as stated previously, the emphasis is on the most commonly used procedures. Thirdly, where reasonable options exist, the protocols are designed to be done by hand or to use the least expensive equipment available. With this in mind, the book does not describe procedures involving electronic imaging instruments or capillary electrophoresis, as such systems are cost-prohibitive to many laboratories.

D. Gersten

ACKNOWLEDGMENTS

Many individuals have made material contributions to the preparation of this volume. Chief among them are my long-time research colleagues, V.J. Hearing, T.M. Phillips and L.V. Rodriguez. Substantial assistance by my academic colleagues, K.E. Bijwaard, J.J. Edwards, S.L. Kamholz, J. Madans, and P.H. Wolf and by my industrial colleagues, L. Beadling, M. Booz, H.A. Daum III, A. Gerstein and M. Saluta is greatly appreciated. I thank J. Hettrick, J. Ray and D. Rickwood for invaluable editorial assistance.

SAFETY

Attention to safety aspects is an integral part of all laboratory procedures and national legislations impose legal requirements on those persons planning or carrying out such procedures. While the authors, editor and publisher believe that the recipes and practical procedures, as set forth in this book, are in accord with current recommendations and practice at the time of publication, they accept no legal responsibility for any errors or omissions, and make no warranty, expressed or implied, with respect to material contained herein. It remains the responsibility of the reader to ensure that the procedures which are followed are carried out in a safe manner and that all necessary safety instructions and national regulations are implemented.

In view of ongoing research, equipment modifications and changes in governmental regulations, the reader is urged to review and evaluate the information provided by the manufacturer, for each reagent, piece of equipment or device, for any changes in the instructions or usage and for added warnings and precautions.

All procedures mentioned within this book must be carried out under conditions of good laboratory practice in accordance with local and national guidelines. Some procedures involve specific hazards, including but not limited to hazards in the following categories:

Chemical. A number of the reagents are known to be carcinogenic, mutagenic, toxic, inflammable, highly reactive or otherwise hazardous. Substances known to be hazardous have been marked with the symbol ⚠ in the list of reagents (but not subsequently) for each protocol, or if they appear as alternatives to the main protocol, the *first time* they appear in the notes. The reader should consult the safety notes on these pages before embarking on any of the procedures covered. This is in no way meant to imply that undesignated chemicals are nonhazardous, and all laboratory chemicals should be handled with extreme caution. Information is not available on the possible hazards of many compounds. The criteria we have generally used for denoting a substance with ⚠ is based upon a hazard level of 2 or more (on a scale 0–4) in any of the categories in the Baker Saf-T-Data™ system used in the material safety data sheets (MSDS) held at the University of Oxford, UK.

These are freely accessible using:
http://joule.pcl.ox.ac.uk/MSDS/.

Other safety infomation on the Internet can be accessed on:
gopher://atlas.chem.utah.edu/11/MSDS
gopher://ginfo.cs.fit.edu:70/lm/safety/msds
http://physchem.ox.ac.uk/MSDS
http://www.fisher1.com/Fischer/Alphabetical Index.html
http://www.pp.orst.edu

You are actively encouraged to check these data sheets to confirm our assignments and for more detailed information on individual hazards; however the author, editor and publisher can accept no responsibility for any material contained in these data sheets. Furthermore, you must always follow the precautions outlined on labels and data sheets provided by individual manufacturers.

Radiation. The use of radioisotopes is subject to legislation and requires permission in most countries. Furthermore, national guidelines for their use and disposal must be rigorously adhered to. The procedures in protocols that use radioisotopes must only be carried out by individuals who have received training in the use of such material using the appropriate facilities, protection and personal monitoring procedures.

Biological. Antibodies, sera and cells (particularly, but not exclusively, those of human and nonhuman primate origin) pose a significant biological hazard. All such materials, whatever their origin, may harbor human pathogens and should be handled as potentially infectious material in accordance with local guidelines. Any recombinant DNA work associated with protocols is likely to require permission from the relevant regulatory body and you must consult your local safety officer before embarking upon this work.

Electrical. Many of the procedures in this book use electrical equipment. Electrophoresis techniques may present particular hazards of this nature.

Lasers. Flow cytometers and certain other types of laboratory equipment contain lasers. Users should ensure they are fully aware of the potential hazards of using such equipment.

I INTRODUCTION TO PROTEIN ELECTROPHORESIS

General information

There are many products and manufacturers whose products are used for electrophoresis and associated technology. The mention of any specific product is not intended as an endorsement of one, or as a negative statement about an unmentioned competing product. Rather, the mention merely reflects the author's familiarity and experience with a particular item.

Protein electrophoresis

For details on the theory of protein electrophoresis, see refs 1–3. Briefly, all charged species will migrate in an electric field, provided that the field is strong enough to propel the mass. Thus, in free solution, the rate at which proteins migrate depends upon the relationship of the protein's charge density to the electric field strength. Electrophoretic mobility is expressed in units of velocity divided by field strength (microns per second/volts per centimeter). For soluble proteins, frictional factors and trajectories due to gravity are of minor consideration. The separations achieved in free solution can be satisfactory in some cases, but can also be seriously limited by convection.

Owing to convection considerations, protein electrophoresis is rarely performed in free solution. Rather, the vast majority of applications use gel matrices, taking advantage of the anticonvective properties offered by semiporous gels. Accordingly, the rate at which soluble proteins move through a gel matrix depends, as above, on charge density and field strength, but also on the relationship of the size of the protein to the size of the pores. Indeed, one major advantage of gel electrophoresis is the ability of the experimenter to achieve superior resolution by adjusting the pore size according to the nature of the protein mixture to be separated. Pore sizing is detailed under gel casting (*Protocol 5*), gradient making (*Protocols 7* and *8*) and stop gradients.

The majority of gel matrices currently in use are polyacrylamide and agarose. Some newer polymers are currently under investigation but they are not in widespread use. Consequently, subsequent discussion is limited to acrylamide and agarose. For a complete discussion of the characteristics and advantages of each of these matrices see refs 4 and 5 for agarose and ref. 6 for polyacrylamide.

The gel format itself is another variable aspect to be considered; gels commonly in use are either cylinders or slabs. Gel casting for slabs and cylinders is described in *Protocols 5–8*. Several factors make slabs the format of choice for most applications.

(i) Electrophoresis of many different specimens in a single slab is frequently more convenient than in a cylindrical format.
(ii) The electrophoretic mobility of different specimens run side-by-side can be readily compared.
(iii) The need for extrusion of fragile gels from a glass cylinder is eliminated.
(iv) Coated glass or plastic backing sheets are commercially available which provide dimensional stability to fragile (thin or very porous) gels (*Protocol 4*).
(v) Overlays containing the necessary substrates and/or cofactors required to demonstrate enzymatic activity can be easily brought into direct contact with gel slabs. Zymography is considered under *Protocols 23, 24* and *25*.
(vi) 'Blotting' of the proteins on to membranes in a manner which faithfully preserves the electrophoretic pattern is possible. *Protocol 21* covers blotting. Indeed, the technique of protein or 'western' blotting from slab gels has become so widespread, that it is the subject of complete textbooks [7–9].

On the other hand, the cylindrical format is sometimes preferred for long runs since lateral diffusion of the bands is limited by the cylinder walls (*Protocol 5c*). The instruments used to prepare and run electrophoresis gels are discussed in Chapters III and IV.

Electrophoretic separations can be achieved in two fundamentally different ways – either by using the intrinsic charge of the proteins, or by imparting an extrinsic charge to the proteins. Separations based on intrinsic charge generally use relatively low ionic-strength electrolytes (for example, Tris-glycine – *Protocol 9*), and attempt to maintain the proteins in their 'native' conformation [10,11]. Another popular approach to separation based on intrinsic charge is isoelectric focusing, in which a pH gradient is established across the electric field. Proteins cease migration at the point in the gradient which corresponds to their isoelectric point (*Protocols 10* and *11*).

It is desirable to perform electrofocusing under 'native' conditions particularly when proteins are to be recovered [12] or used for zymography [13,14]. Alternatively, focusing of denatured proteins is sometimes used when resolution of their component subunits is desired. This is most frequent in two-dimensional experiments (*Protocol 13*). Still another application using denatured proteins is 'nonequilibrium pH gradient electrofocusing,' carried out as a part of two-dimensional separations [15].

The most common form of electrophoresis currently practiced is SDS–polyacrylamide gel electrophoresis (*Protocol 12*). SDS-denatured proteins rarely retain functional properties such as enzymatic activity. If obtaining functional protein is the object of the experiment, steps must therefore be taken, following the electrophoretic separation, to restore the native conformation to the protein. Renaturation options are discussed in Chapter VI.

Two-dimensional electrophoresis, combining electrofocusing in the first dimension and pore gradient SDS–electrophoresis in the second dimension is a popular technique for the separation of complicated mixtures of proteins. Here, the molecules are separated at right angles on the basis of two unrelated parameters, isoelectric point and SDS–molecular weight. Consequently, the amount of data that can be obtained from the analysis of the protein spots in a single gel pattern is enormous (*Protocol 13*).

Proteins, in the amounts used for gel electrophoresis are, apart from the few colored proteins, invisible to the naked eye. Therefore, following gel electrophoresis, it is necessary to fix the proteins in position and then detect their location in the gel by staining. The most common alternative to staining, is to label the proteins with radionuclides. The options for protein detection are covered in Chapter V.

In many cases, it is not sufficient merely to resolve and analyze the various components of a mixture of proteins, or to assess the purity of a particular preparation. It is necessary to identify the proteins as well. There are two major strategies which are most often used to assign identity to individual protein bands or spots in the gel. These are immunochemical techniques (western blotting – *Protocol 21*, immunosubtraction – *Protocol 22* and immunoelectrophoresis – *Protocols 15* and *16*), and zymographic techniques (*Protocols 23, 24* and *25*).

Finally, it is frequently necessary to maintain a permanent record of the gel. This can be accomplished either by preserving the gel itself (*Protocol 27*), or by taking a photograph or making a photocopy (*Protocol 26*).

References

1. Dunn, M.J. (1993) *Gel Electrophoresis: Proteins.* BIOS Scientific Publishers, Oxford.
2. Mosher, R.A., Saville, D.A. and Thormann, W. (1992) *The Dynamics of Electrophoresis.* VCH, Weinheim.
3. Westermeier, R. (1993) *Electrophoresis in Practice.* VCH, Weinheim.
4. *The Agarose Monograph*, 4th edn (1988) FMC Bioproducts, FMC Corp, Rockland, ME, USA.
5. Stellwagen, J. and Stellwagen, N.C. (1994) *Biopolymers* **34**, 187–201.
6. Righetti, P.G. (1981) in *Electrophoresis '81* (R.C. Allen and P. Arnaud, eds), pp. 3–16. Walter de Gruyter, Berlin.
7. Baldo, B.A. and Tovey, E.R. (eds) (1989) *Protein Blotting.* Karger, Basel.
8. Dunbar, B.S. (1994) *Protein Blotting.* IRL Press, Oxford.
9. Bjerrum, O.J. and Heegaard, N. (1988) *Handbook of Immunoblotting of Proteins.* CRC Press, Boca Raton, FL.
10. Davis, B.J. (1964) *Ann. N.Y. Acad. Sci.* **121**, 404–427.
11. Ornstein, L. (1964) *Ann. N.Y. Acad. Sci.* **121**, 321–349.
12. Righetti, P.G., Faupel, M. and Wenisch, E. (1992) *Adv. Electrophor.* **5**, 159–200.
13. Gabriel, O. and Gersten, D.M. (1992) *Analyt. Biochem.* **203**, 1–21.
14. Gersten, D.M. and Gabriel, O. (1992) *Analyt. Biochem.* **203**, 181–186.
15. O'Farrell, P.Z., Goodman, H.M. and O'Farrell, P.H. (1977) *Cell,* **12**, 1133–1142.

II SAMPLE PREPARATION

In order to achieve optimal separation of proteins in electrophoresis gels, the sample must be fully solubilized and an accurate measurement of the amount of protein loaded must be made. These two points are crucial if reliable and reproducible results are to be obtained.

Protein loading (see *Table 1*)

For analytical electrophoresis, three factors should be taken into account when considering the amount of protein loaded on to an electrophoresis gel. These are: thickness of the gel, number of bands or spots contained in the sample and, primarily, the detection method. Detection methods are discussed in Chapter V. For preparative applications, the size of the gel is the primary concern. Since the proteins are recovered in solution, they can be detected spectrophotometrically or following subsequent electrophoresis of small aliquots of the fractions in an analytical gel.

Protein solubilization

Solubilization under native conditions is performed when: (i) the nature of native proteins is to be studied,(ii) for some preparative applications where the protein is to be recovered, (iii) protein function is to be assessed following electrophoresis – for example, zymographic applications and protein interaction studies. Solubilization under denaturing conditions is performed when the subunit structure of proteins is to be assessed or when they remain insoluble under milder conditions. This can be done with or without reduction of disulfide bridges.

As the choice of solubilization technique depends on the nature of the protein and object of the experiment, no universal method is available. Each of the gel techniques given in Chapter IV has its own requirements and the reader will find it helpful to consider the relevant sections before preparing the sample.

Table 1. Protein loading guide[a]

	Protein loading band			
	Soluble stains	Colloidal stains	Silver stains	Radioactive detection
Vertical slab, 0.5–1.5 mm	0.5–1 μg	0.2–0.4 μg	80–160 ng	100–1000 d.p.m.[b] 1000–10 000 d.p.m.[c]
Vertical slab, 1.5–3 mm	1–5 μg	0.4–2.0 μg	160–800 ng	100–1000 d.p.m.[b] 1000–10 000 d.p.m.[c]
Vertical slab, stop gradient	0.1–1 μg	40–400 ng	20–160 ng	100–1000 d.p.m.[b] 1000–10 000 d.p.m.[c]
SDS horizontal slab, 0.5–1.0 mm[d]	0.5–3 μg	0.2–1.2 μg	80–400 ng	100–1000 d.p.m.[b] 1000–10 000 d.p.m.[c]
IEF horizontal slab, 0.5–1.0 mm[d]	0.5–1 μg	0.2–0.4 μg	80–160 ng	100–1000 d.p.m.[b] 1000–10 000 d.p.m.[c]
IEF ultrathin slab, <0.2 mm	<1 μg	<1 μg	<1 μg	n/a[e]
Cylindrical, 5 mm diameter	2–5 μg	n/a	n/a	100–1000 d.p.m.[b]
Cylindrical, 2 mm diameter, 1st dimension of 2D	500 μg	200 μg	80 μg	10^5–5×10^6 d.p.m.[c]
Slab, 1st dimension of 2D[d]	500 μg	200 μg	80 μg	10^5–5×10^6 d.p.m.[c]
Preparative	50 mg crude extract – 1 mg/ band of interest[f]	n/a	n/a	n/a
Immunoelectrophoresis	5–20 μg	See note g	See note g	1000–10 000 d.p.m.[c]

[a]Approximate ranges of protein loads for various gel formats, thicknesses and detection methods. Actual loads will vary depending on the nature of the sample and purpose of the experiment. [b]Radioactivity detected using electronic counting methods. (Discussed in Chapter VII.) [c]Radioactivity detected by autoradiography or fluorography. (Discussed in *Protocol 20*.) [d]Detection sensitivity of colloidal and silver stains is reduced in slab gels which are not removed from their backing. Backing supports for thin or fragile slab gels are discussed in *Protocol 4*. [e]Radioactive detection is rarely performed in ultrathin gels. The resolution advantages resulting from 'tightness' of bands (discussed in *Protocol 10c*) is negated by the 'fuzziness' of radioactive detection. [f]For two adjacent bands separated by one band width, a maximum load of 1–2 mg/cm^2 of cross-sectional area can be applied [1]. [g]Use of colloidal and silver stains can frequently give an unacceptably high background, due to residual antibodies in the gel.

In general, salt solutions of higher rather than lower ionic strength afford better protein solubility. However, low ionic strength is frequently advantageous for optimal electrophoretic separations. As a result, low concentrations of detergents are usually added to proteins whose solubility is incomplete at low ionic strength. Nonionic (e.g. Triton X-100) and zwitterionic (e.g. CHAPS; 3-[3-(chloramidopropyl) dimethylammonio]-1-propane sulfonate) detergents are useful for non-denaturing conditions and anionic detergents (e.g. SDS) are used for denaturing conditions. A good discussion on the use of detergents can be found elsewhere [2].

Solubilization under native conditions

As stated previously, no universal methods are appropriate for all protein sources and all gel techniques. Consequently, some general considerations and guidelines are given.

First, we will consider buffer systems and their conductivity. When continuous buffer systems are used, the sample is applied in the same buffer that is used in the gel and buffer tank; explanation of continuous and discontinuous buffer systems is given in the Introduction to Chapter IV. When discontinuous systems are used, conductivity (i.e. ionic strength of the buffer) of the sample should be approximately the same as that of the stacking gel. Conductivity discontinuities result when the sample has a different ionic strength than the gel. These are less of a problem in vertical than horizontal systems; vertical and horizontal systems are explained in the Introduction to Chapter III. For horizontal systems, the sample should be in a buffer of equal or slightly greater conductivity than the gel surface to which it is applied, otherwise current will be shunted away from the sample and the proteins will not be drawn into the gel.

If the salt concentration in the protein preparation is too high, the following methods can be employed:

(i) Dialyze the sample against sample buffer. For most experiments, a dialysis tube with a nominal molecular weight cut-off of 12 000 Da is appropriate.

(ii) Desalt the protein on a small column equilibrated in sample buffer. BioGel P-2 (BioRad) and Sephadex G-10 (Pharmacia) are good matrices for this purpose; however check for protein losses on the column.
(iii) Precipitate the protein with acetone – mix the sample with at least five volumes of freezer-cold (−20°C) acetone, incubate in the freezer for 15–20 min and harvest the precipitate by centrifugation for 5 min at 10 000 *g*. Wash the precipitate three times in freezer-cold acetone then decant and dry the precipitate in an airstream at room temperature. Dissolve the dried precipitate in sample buffer.
(iv) If other methods fail, then try precipitating the protein with trichloroacetic acid (TCA) – mix the protein preparation with an equal volume of 20% (w/v) ice-cold TCA solution. Incubate 30 min at 4°C and harvest the precipitate by centrifugation for 5 min at 10 000 *g*. Remove excess TCA by washing the precipitate at least three times with a 50/50 mixture of ethanol/ethyl ether then redissolve in sample buffer; make sure that the sample is completely solubilized. Incomplete removal of TCA may exceed the buffering capacity of the sample buffer. Test the final pH of sample immediately before loading on to the gel by spotting 2–5 μl on a piece of pH paper and adjust the pH if necessary.

If the protein concentration in the sample is too dilute for the detection method contemplated:

(i) consider using a more sensitive detection method,
(ii) concentrate the sample using either lyophilization, rotary evaporation (e.g. Savant Speedvac), dialysis against dry carbowax (polyethylene glycol 8000) or Sephadex (G-200) or by ultrafiltration (e.g. Amicon PM-10 filter).

Note that for each of these options, careful attention must be paid to the final salt concentration.

If the sample contains too many different proteins, so that the relative abundance of the protein of interest is too low, the sample can be fractionated before electrophoresis by several different methods:

(i) *Gel filtration.* Fractionation on the basis of size using gel filtration is particularly useful when the subsequent electrophoretic separation will be performed on the basis of charge rather than size [3]. (Explanation of size- vs. charge-based separation is given in the Introduction to Chapter IV.)

(ii) *Affinity chromatography.* Cibacron Blue conjugated to beaded agarose is a frequently used affinity matrix for selective adsorption of serum albumin from whole serum and nucleotide-binding proteins. Prepare a 1 ml bed volume column of Blue Sepharose (Pharmacia), swollen in column buffer (0.1 M NaCl, 0.05 M Tris-HCl, pH 8.0). Apply 100 μl of whole serum to the top of the bed and elute with 3 ml of column buffer. (The swollen matrix can be stored at 4°C in column buffer containing 1 mg/ml NaN_3.) Dye bound matrices have been used for a wide range of applications.

(iii) *Precipitation with ammonium sulfate.* For selective precipitation add solid ammonium sulfate to a final concentration of 75% (w/v) and incubate at 4°C for 30 min. Centrifuge at 18 000 *g* for 10 min, then dissolve the pellet in buffer (e.g. 0.1 M Tris-HCl, pH 7.2). Dialyze against sample buffer. Before use, treat the dialysis tubing by boiling for 20 min in 0.1 M $NaHCO_3$. If the protein concentration in the final dialysate will be too low, dialyze against 1/10 strength sample buffer and concentrate the final dialysate 10-fold.

(iv) *Immunoprecipitation.* This is frequently used for separations of radioactive proteins metabolically labeled in cell culture. To accomplish this, wash the cells in buffer or serum-free medium, then resuspend and incubate them in balanced salt solution or culture medium containing radioisotope at a concentration of at least 10^8 d.p.m./ml. Lyse the cells in hypotonic solution or by freezing and thawing them, add CHAPS to a final concentration of 2% (w/v), and centrifuge at 30 000 *g* for 30 min. To 100 μl of supernatant, add 5 μl of antiserum and incubate at 4°C for 1 h. The volume of antiserum will vary with the titer of specific antibodies it contains and this must be determined experimentally. If sufficient radioactivity is not precipitated within 1 h then leave overnight; longer incubations are usually not worthwhile. Prepare a 20% slurry of Protein A-Sepharose in Tris-buffered saline (TBS) and add 80 μl. Incubate at 4°C for 1 h with end-over-end rotation. Centrifuge the sample at 12 000 *g* for 1 min, decant the supernatant, wash the beads three times in 100 μl of TBS and recentrifuge. Discard the supernatant and elute the bound radioactivity from the beads by adding 100 μl of elution buffer. (Some antigens can be dislodged from the Protein-A Sepharose matrix directly by sample buffer.)

To determine if the sample has particulates in it, its turbidity can be assessed using a spectrophotometer at 600–650 nm. If particulate matter is present several options are available:

(i) The sample can be filtered through a 0.22 μm membrane filter, pre-wet with sample buffer.
(ii) The sample can be centrifuged in a bench top microcentrifuge at 12 000 *g* or a refrigerated centrifuge at higher speeds.
(iii) Detergents can be added – the nonionic detergent NP-40 is used for protein extraction from cells and tissue specimens at concentrations ranging between 0.05 and 1% (w/v). When present during the electrophoresis, it is rarely used at concentrations higher than 0.1%. CHAPS is usually used for isoelectric focusing at concentrations of about 2% (w/v). A typical detergent-containing sample buffer for isoelectric focusing consists of 2% CHAPS, 0.5% (w/v) Ampholytes, pH 5–7. (See Chapter IV for a discussion of ampholytes.)

Inhibitors of proteolytic enzymes are included in some sample preparation procedures if degradation of the sample is likely. Two of the most common are aprotinin (1 μg/ml) and phenylmethylsulfonyl fluoride (PMSF). Aprotinin can be prepared 100× in sample buffer and stored frozen. Prepare PMSF as 100 mM (17.4 μg/ml) in dried isopropanol (store at −20°C) and, immediately before use, dilute into the sample to a final concentration of 1 mM in buffer.

Solubilization under denaturing conditions
There are many possible approaches to solubilization of proteins under denaturing conditions. The two most commonly used in electrophoretic applications are ionic detergents (primarily SDS) for breaking hydrophobic interactions and polar solvents (primarily urea) for breaking hydrogen bonds. Agents which reduce inter- and intra-chain disulfide bridges, such as 2-mercaptoethanol and dithiothreitol, are often used in combination with SDS and urea.

To prepare samples for SDS–polyacrylamide gel electrophoresis, mix the sample with an equal volume of 2× SDS sample buffer (0.98 g Tris-HCl, 2.0 g SDS, 7.5 ml glycerol, 5 ml 2-mercaptoethanol. Adjust pH to pH 6.8, make up to 50 ml with distilled water) and incubate in a boiling water bath for 3 min. Samples prepared in this sample buffer can be stored indefinitely at −20°C. They should be heated immediately prior to electrophoresis. If SDS–polyacrylamide gel electrophoresis without reduction of disulfide bridges is desired, omit the mercaptoethanol from the SDS sample buffer.

When samples enriched using immunoprecipitation (see Preparation under native conditions) are subsequently to be analyzed by SDS electrophoresis, SDS sample buffer is frequently used in place of elution buffer to displace the antigen from the Protein A-Sepharose matrix.

To prepare serum samples for two-dimensional electrophoresis, mix 10 μl of serum with 30 μl of SDS-CHES solution (100 mg CHES [Calbiochem (full name is cyclohexylaminoethane sulfonic acid)], in 7 ml distilled water. Adjust pH to pH 9.5 with 0.5 M NaOH then add 200 mg SDS and 0.5 ml of 2-mercaptoethanol or 100 mg of dithiothreitol (final vol. 8 ml) (SDS-CHES solution can be stored at room temperature as long as the pH is maintained) and heat in a boiling water bath for 3 min. Allow to cool to room temperature, then add 80 μl of urea–NP-40 solution (5.4 g recrystallized urea, 0.4 ml NP-40, 0.2 ml Ampholytes pH 3–10, 0.5 ml 2-mercaptoethanol. Make up to 10 ml in distilled water).

Most two-dimensional techniques use isoelectric focusing (charge-based separation) in the first dimension and SDS electrophoresis (size-based) in the second dimension. For a description of charge- and size-based separations, see the Introduction to Chapter IV. Proteins to be separated on the basis of charge should never be heated in the presence of urea as this causes carbamylation, due to the formation of cyanate ions, resulting in the creation of charge heterogeneity. When urea is present in the sample buffer, Tris base should be included also, as it has an amino group able to mop up the cyanate ions produced from the decomposition of urea. SDS swamps out the protein's normal intrinsic charge and so charge-based separations of SDS-solubilized protein must be performed in the presence of NP-40, which will displace SDS from the protein.

Urea is most effective as a denaturant above concentrations of about 7 M. If the sample buffer includes urea below this concentration, it should be present at a higher concentration in the gel (see *Protocol 13*).

For experiments where solid tissue samples are used, homogenize the sample directly in SDS-CHES solution, remove any particulates by centrifugation then heat, cool and mix with urea–NP-40 solution as described previously. Serum or tissue samples can be stored indefinitely at −20°C in SDS-CHES solution. Heat, cool to room temperature and mix with urea–NP-40 solution immediately before electrophoresis.

Estimation of proteins

In electrophoresis gels, there is an appreciable increase in the resolution of proteins and peptides as the amount of sample applied to the gel decreases [4–6]. The amount of sample loaded, of course, depends primarily on the detection method. The detection of the proteins is discussed in Chapter V.

For most experiments, where the detection is to be done by staining, colorimetric methods such as those described in *Protocols 1* and *2* are used. When detection by more sensitive methods is to be used and less protein is loaded into the gel, fluorometric [7] or amplified methods [8] are available.

In analytical experiments where the purpose is to resolve the components of a mixture, the strategy is to load as much protein as possible without allowing the bands or spots to merge. In analytical experiments where the object is to assess the purity of a particular preparation, the strategy is to deliberately 'overload' the gel in order to detect the presence of any potential contaminants. Regardless of the object of the experiment, however, reliable and reproducible results require the accurate measurement of the protein sample.

Methods available

The Bradford technique (see *Protocol 1a*)
This method for estimation of proteins [9] makes use of the fact that the dye, Coomassie Brilliant Blue G-250 (CI 42655), undergoes a spectral shift in its absorption maximum from 465 to 595 nm when bound to the free amino groups of proteins. Its advantage lies in the fact that it is a rapid, one-step reaction with a fairly wide linear range.
Problems: Not appropriate for some applications,such as first dimension of two-dimensional experiments or those where carrier ampholytes are used to solubilize proteins. The modified Bradford technique (*Protocol 1b*) is used in the presence of ampholytes, organic detergents and urea.

The Lowry method
Protein estimation by the traditional technique of Lowry *et al.* [10] is one of the most widely used methods in biomedical science.
Problems: The color fades with time, the reaction takes about 40 min to complete and a wide range of substances interfere with the reaction.

Bicinchoninic acid
Like the Lowry reaction, protein estimation with bicinchoninic acid (BCA) measures the reduction of Cu^{2+} to Cu^{+} [11], but it is easier. The advantage of the BCA over the Lowry method is that the purple color does not fade with time, the reaction is not photosensitive and

References

1. Rodbard, D., Chrambach, A. and Weiss, G.H. (1974) in *Electrophoresis and Isoelectric Focussing in Polyacrylamide Gel* (R.C. Allen, ed.), p. 63. De Gruyter, Berlin.
2. Dunbar, B.S. (1987) *Two-dimensional Gel Electrophoresis and Immunological Techniques*, pp. 47–65. Plenum Press, New York.
3. MacGillivray, A.J., Johnston, C., MacFarlane, R. and Rickwood, D. (1978) *Biochem. J.* **175**, 35–46.
4. Radola, B.J. (1980) *Electrophoresis*, **1**, 43–56.
5. Allen, R.C. (1980) *Electrophoresis*, **1**, 32–37.
6. Kling, H., Sawatski, G. and Geis, W. (1980) *Analyt. Biochem.* **174**, 589–592.
7. Ohkura, Y., Kai, M. and Nota, H. (1994) *J. Chromatog. B.* **659**, 85–107.
8. Butcher, E.C. and Lowry, O.H. (1976) *Analyt. Biochem.* **76**, 502–523.
9. Bradford, M.M. (1976) *Analyt. Biochem.* **72**, 248–254.
10. Lowry, O.H., Rosebrough, N.J., Farr, A.L. and Randall, R.J. (1951) *J. Biol. Chem.* **193**, 265–275.
11. Smith, P.K., Krohn, R.I., Hermanson, G.T. *et al.* (1985) *Analyt. Biochem.* **150**, 76–85.
12. Brown, R.E., Jarvis, K.L. and Hyland, K.J. (1989) *Analyt. Biochem.* **180**, 136–139.
13. Morrissey, T.B. and Woltering, E.A. (1989) *J. Surg. Res.* **47**, 273–275.
14. Gates, R.E. (1991) *Analyt. Biochem.* **196**, 290–295.

fewer substances interfere [12].

Problems: Substances which influence the reduction of Cu^{2+} can interfere with the assay.

Amido Black filter assay (see *Protocol 2*)

The estimation of proteins on membrane or fiberglass filters eliminates some of the difficulty with interfering small molecules. In these assays, protein is precipitated on to the filters using trichloroacetic acid. The filter is then stained and destained with a protein stain. (For a discussion of protein stains see Chapter V.) The stain is then eluted from the filter and binding is quantified spectrophotometrically [15,16].

15. Schaffner, W. and Weissman, C. (1973) *Analyt. Biochem.* **56**, 502–514.
16. McKnight, G.S.(1977) *Analyt. Biochem.* **78**, 86–92.
17. Neuhoff, V., Stamm, R. and Eibl, H. (1985) *Electrophoresis*, **6**, 427–448.
18. Bensadoun, A. and Weinstein, D. *Analyt. Biochem.* **70**, 241–250.
19. Akins, R.E. and Tuan, R.S. (1992) *BioTechniques*, **12**, 496–499.

Protocols provided

1a. *Protein estimation by the Bradford technique*

1b. *Modified Bradford technique for protein determination in the presence of carrier ampholytes*

2. *Amido Black filter assay*

15

Sample preparation

Protocol 1a. **Protein estimation by the Bradford technique**

Reagents

10 mg Purified Coomassie Brilliant Blue G-250 (also called Serva Blue G-250 and Xylene Brilliant Cyanine G) (Serva Fine Biochemicals) in 5 ml absolute ethanol ⚠ ①

85% H_3PO_4 ⚠

Ovalbumin

Equipment

Disposable cuvettes

Filter paper (Whatman No. 1 or equivalent)

Spectrophotometer

Test tubes

Procedure

1 To the Coomassie Brilliant Blue G-250 solution, add 10 ml of 85% H_3PO_4.

2 Add 185 ml distilled water and filter. ②

3 Prepare a standard curve of ovalbumin by adding 50 μl of the following concentrations (0, 0.125, 0.25, 0.5, 0.75, 1.0 mg/ml), to 5 ml of the above solution. Add 50 μl of sample to 5 ml of above solution. Samples with higher concentrations can be diluted. ③

4 Incubate for 2 min, but not longer than 1 h, at room temperature. [1]

5 Read optical density at 595 nm. ④

Notes

This procedure will take approximately 15 min.

① Coomassie Brilliant Blue G-250 from different suppliers varies in purity from about 35–95%. Serva claims 95% purity. If difficulty is encountered with the sensitivity, purify the stain by precipitation in ammonium sulfate or ammonium acetate (see note 2 to *Protocol 18* or ref.17).

② Coomassie Brilliant Blue rarely dissolves completely, even after purification. Filtration immediately before use is advised.

③ Since the reaction measures amino groups, very basic proteins such as immunoglobulins will be overestimated and acidic proteins underestimated. Accordingly, substitute Cohn fraction II IgG or Cohn fraction V BSA for ovalbumin in the standard curve when basic or acidic proteins (respectively) are to be estimated.

④ The stain sticks to the walls of spectrophotometer cuvettes making them difficult to clean between uses. The use of disposable plastic cuvettes is recommended.

Protocol 1b. **Modified Bradford technique for protein determination in the presence of carrier ampholytes**

Reagents

10 mg Purified Coomassie Brilliant Blue G-250 (also called Serva Blue G-250 and Xylene Brilliant Cyanine G) (Serva Fine Biochemicals) in 5 ml absolute ethanol ⚠ ①
0.1 M HCl ⚠
85% H_3PO_4 ⚠
Ovalbumin

Equipment

Disposable cuvettes
Filter paper (Whatman No. 1 or equivalent)
Spectrophotometer
Test tubes

Procedure

1 Perform steps 1–3 of *Protocol 1a*.

2 Construct a standard curve by mixing samples containing 1–50 μg of ovalbumin in 10 μl of urea mix (the two-dimensional electrophoresis sample buffer containing 9 M urea, 2% carrier ampholytes, 4% NP-40 and 2-mercaptoethanol). Then add 10 μl of 0.1 M HCl. Acidify 10 μl of the sample in urea mix in the same way. ②

3 Perform steps 4 and 5 of *Protocol 1a*.

Notes

This procedure will take approximately 15 min.

① See note 1 of *Protocol 1a*.

② Less than 50 μl are usually loaded into the first dimension of a two-dimensional experiment (see *Protocol 13*).

Protocol 2. **Amido Black filter assay**

Reagents

Bovine serum albumin
Destaining solution – methanol:acetic acid:distilled water 90:2:8 (v/v/v) ⚠
Eluant – 0.05 M EDTA, 25 mM NaOH in 50% (v/v) ethanol ⚠
Staining solution – 1% (w/v) Amido Black 10B (CI 20470) in methanol:acetic acid:distilled water 45:10:45 (v/v/v)
60% (w/v) Trichloroacetic acid (TCA) ⚠
Tris–SDS solution – 1% (w/v) SDS in 1 M Tris-HCl, pH 7.5

Equipment

Cuvettes
Glass or plastic filtering assembly
Nitrocellulose membrane filters (Millipore or equivalent, 0.45 μm pore size)
Spectrophotometer
Test tubes
Vortex mixer

Procedure

1 Prepare a standard curve by adding 0–150 μg of fraction V bovine serum albumin to 270 μl of water. Dilute the samples to 270 μl. ① ②
2 Add 30 μl of Tris-SDS solution then 600 μl of 60% TCA and allow to stand for 15 min.
3 Filter, rinse the tubes twice with 2 ml of 6% TCA and use the rinse to wash the filter. ③ ④
4 Remove the filter from the filtration apparatus and incubate the filter for 2–3 min in staining solution, with agitation.

Notes

This procedure will take approximately 30 min.

① Potassium ions must be excluded from this assay as they precipitate dodecyl sulfate.
② Caution must be exercised when using filter assays to estimate protein in samples containing carrier ampholytes. The ampholytes, which will stain with Amido Black, are sometimes difficult to remove from the filter.
③ Millipore filters should be pre-wet by floating the filter on water. Immersing them will cause air to be trapped in the interior of the filter. Filters may be labeled using a soft

5 Rinse for 30 sec in distilled water then three times for 1 min each in destaining solution, then once more in distilled water.

6 Blot excess water from the filters then incubate with 0.6 ml eluant for 10 min with occasional vortexing. ⑤

7 Add 0.9 ml of eluant, decant and read the optical density at 630 nm. [1]

lead pencil before use.

④ The sample volume that can be applied to a 25 mm filter disc varies with salt concentration and from protein to protein. The protein must remain in a concentrated spot of less than 10 mm diameter.

⑤ The sensitivity can be increased by using proportionally less eluant.

Pause point

[1] Once stain is eluted from the filter, it can be left indefinitely before reading.

III GEL PREPARATION

Introduction

When proteins migrate through a semiporous gel matrix, the rate of migration is influenced by the relationship of the size of the protein molecules to the size of the gel pores. Thus the superior resolving power of gel electrophoresis is, to a large extent, attributable to the use of gels whose pore size is tailored by the experimenter to the mixture of proteins to be separated (see refs 1–3 in Chapter I).

Methods available

Gel matrices available

Polyacrylamide gels

For polyacrylamide gels, the average pore size is determined by the acrylamide monomer concentration [%T for total monomer concentration (w/v)] and the concentration of cross linker (%C for percentage of the mass of monomer which is cross linker). The most frequently used cross linker is bisacrylamide (full name *N,N′*-methylene bisacrylamide). For example, a solution of 100 ml which has 9.7 g of acrylamide monomer and 0.3 g of bisacrylamide cross linker would, when polymerized, be characterized as a gel of 10%T, 3%C.

Acrylamide monomer, when mixed with initiators (usually ammonium persulfate or riboflavin) and an accelerator (usually TEMED;

References

1. Kiehm, D.J. and Tae, H.J. (1987) *J. Biol. Chem.* **252**, 8524–8531.
2. Schagger, H. and von Jagow, G. (1987) *Analyt. Biochem.* **166**, 368–379.
3. Judd, R.C. (1994) *Meth. Mol. Biol.* **32**, 49–57.
4. Okajima, T., Tanabe, T. and Yasuda, T. (1993) *Analyt. Biochem.* **211**, 295–300.
5. Grierson, D. (1990) in *Gel Electrophoresis: a Practical Approach* (D. Rickwood and B.D. Hames, eds), pp. 1–49. IRL Press, Oxford.
6. Westermeier, R. (1993) *Electrophoresis in Practice*, p. 57. VCH, Weinheim.
7. Allen, R.C. and Saravis, C.A. (1987) *Biotechniques*, **5**, 248–252.
8. Fawcett, J. and Chrambach, J. (1986) *Electrophoresis*, **7**,

N,N,N′,N′-tetramethylethylenediamine) will form a linear polymer, having the consistency of a viscous liquid. The incorporation of a cross linker into the linear polymer joins the linear polymers together, side-to-side, to form a three-dimensional mesh. Average pore size is thus determined by the number of linear polymers per unit volume (a function of %T) and the frequency of interchain bridges, determined by the extent to which bisacrylamide is substituted for acrylamide (%C).

In common use, polyacrylamide gels range in concentration from about 4%T to 40%T. Below about 3%T polyacrylamide gels have little dimensional strength regardless of the %C. Therefore, very large proteins or aggregates whose resolution requires larger pores are usually separated using agarose gels or hybrid gels (also called composite gels). Hybrid gels are a mixture of agarose and polyacrylamide [1]. At the other extreme, acrylamide concentrations above about 40%T are rarely useful because the pores are too small to permit migration of all but small peptides. Methods for the resolution of small peptides have been described [2–4].

When using polyacrylamide gels, only the highest quality acrylamide monomer should be used. This is particularly important in zymographic applications, since the loss and/or inhibition of enzymatic activity due to the presence of contaminants in the gel is unpredictable. Consequently, the author recommends recrystallizing

260–265.

9. Allen, R.C., Budowle, B. and Reeder D.J. (1993) *Appl. Theor. Electrophor.* **3**, 173–181.
10. Gersten, D.M., Kimball, H. and Bijwaard, K.E. (1991) *Analyt. Biochem.* **197**, 59–64.
11. Bjellqvist, B., Basse, B., Olsen, E. and Celis, J.E. (1995) *Electrophoresis*, **15**, 529.
12. Allen, R.C., Budowle, B., Saravis, C.A. and Lack, P.M. (1986) *Acta Histochem. Cytochem.* **19**, 637–645.
13. Allen, R.C. (1987) *ACS Symp. Ser.* **335**, 117.
14. Dossi, G., Celentano, F., Gianazza, E. and Righetti, P.G. (1983) *J. Biochem. Biophys. Meth.* **7**, 123–142.
15. Ruchel, R. and Gross, J.(1979) *Analyt. Biochem.* **92**, 91–98.
16. Dhamankar, V.S., Choudhury, M.D. and Vartak, H.G. (1986) *Analyt. Biochem.* **157**, 289–290.
17. Radola, B.J. (1980) *Electrophoresis*, **1**, 43.
18. Frey, M.D., Kinzkofer, A., Atta, M.B. and Radola, B.J. (1986) *Electrophoresis*, **7**, 728–40.
19. Allen, R.C. and Lack, P.M. (1987) in *New Directions in Electrophoretic Methods* (J.W. Jorgenson and M. Phillips, eds), pp. 117–131. ACS Symposium Series, Washington, DC.
20. Domingo, A. (1990) *Analyt. Biochem.* **189**, 88–90.

the acrylamide using chloroform, or bisacrylamide using acetone, before use [5].

Agarose gels (see *Protocol 5d*)
Agaroses are long-chain polysaccharides. They are generally characterized by their melting point and electroendosmosis properties.

Agarose gels are commonly described by their % (w/v); gels of about 0.5–2% are in common use in protein electrophoresis. An agarose gel of 1% has a far greater dimensional stability than a 4%T, 4%C polyacrylamide gel despite a pore size about 10 times larger. Hence, agarose is the matrix of choice for separations in which very large molecules and aggregates are involved. It is also the medium of choice for immunoelectrophoresis in which subsequent diffusion of antibodies is required. It is also used to strengthen hybrid gels.

Gel casting formats

Where possible, it is recommended that polyacrylamide slab gels, especially those run in the horizontal format, be covalently anchored to a plastic backing sheet or on to a silanized glass plate. Agarose gels can also be anchored to a plastic backing sheet or glass support. This allows the use of electrophoresis gels having relatively large pores and therefore poor physical strength. The advantages of backed gels are many, including:

Protocols provided

3. *Preparation of glassware*
4. *Preparation of backing sheets*
5a. *Gel casting for vertical polyacrylamide slab gels*
5b. *Gel casting for horizontal polyacrylamide slab gels*
5c. *Casting cylindrical polyacrylamide gels*
5d. *Agarose slab gel electrophoresis*
6. *Rehydratable gels*
7. *Preparation of gradient gels using gradient markers*
8. *Preparation of gradients by the tilting method*

(i) Tearing of fragile gels is eliminated.
(ii) Backed gels retain their rectangular shape. Unbacked gradient gels, when fixed and stained, will assume a trapezoidal shape, which makes the comparison of samples in different lanes difficult.
(iii) In applications where molecular sieving is not the object, gels with larger pore sizes reduce separation time considerably.
(iv) In zymography and protein interaction studies, large-pore gels are advantageous because they more readily permit the rapid diffusion of small molecules into the gel following electrophoresis.
(v) Gels can be preserved and stored directly on the backing sheet.

The disadvantages of backed gels are that fixing, staining and destaining times are increased since these reagents enter the gel from only one surface. More importantly, electroblotting, which requires both surfaces to be accessible, is not possible. In addition the transfers for capillary blotting are incomplete because buffer reservoirs cannot be used. However, an instrument now exists for cleanly slicing a gel off its backing sheet [6].

Types of gel apparatus

Polyacrylamide gel electrophoresis can be performed in horizontal gel slabs, vertical slabs and vertical cylinders, using a variety of equipment. Glassware used for electrophoresis should always be scrupulously clean (see *Protocol 3*). The most common arrangement is to use vertical slabs, especially for discontinuous gels.

Electrophoresis in vertical slabs (see *Protocol 5a*)
When voltage is applied across a gel, the gel heats up. The heating can become a serious problem which, in its extreme, results in boiling and melting of the gel. Thus the voltage which can be applied to any system (therefore the speed at which the separation can be performed) is limited by heating considerations. Electrical resistance of the gel depends on the buffer used and decreases as the cross-sectional area of the gel increases. Accordingly, increasing the thickness of the gel will produce less heat or permit higher applied voltage for the same amount of heat production in the gel. A thicker gel, however, has only limited benefit since heat retention by a thicker gel is greater.

The problem of heating is best addressed by cooling. For vertical slab gels, the arrangements which are most popular are those in which the plates are immersed almost totally in the lower electrolyte buffer reservoir. A heat exchanger which circulates cold water is also included in the lower reservoir. This is shown in *Figure 1*. Vertical minigel systems are also available with a similar design or a design in which one side of the gel is immersed in upper electrolyte and the other in lower electrolyte (*Figure 2*). Minigel systems are small

Figure 1. Vertical slot gel instrument.

enough that the entire apparatus can be placed in a cooled bath.

It is advantageous to maintain uniform temperature throughout the gel. The two most important reasons are ionization and pore size. Firstly, for separations which rely on the protein's intrinsic charge, it should be remembered that the ionization constants of the amino acids which determine the net charge are temperature dependent. Thus, if the temperature is uneven, the same protein in two different lanes of the same gel can have different electrophoretic mobilities. Secondly, heating causes pore sizes to enlarge somewhat and so a protein will migrate faster through a hotter part of the gel.

Electrophoresis in horizontal slabs (see *Protocol 5b*)

For certain applications, electrophoresis in a horizontal format is the method of choice. Horizontal electrophoresis has generally been restricted to those applications not requiring a stacking gel (e.g. isoelectric focusing and immunoelectrophoresis). This restriction has recently been removed to some extent with the advent of 'seamless' discontinuous gels (see p. 91). The two main advantages offered by the horizontal format are improved cooling and the ability to load the sample anywhere in the gel. Two types of cooling platen are in common use. The least expensive is a glass platen through which chilled water is circulated. The second is a slab of beryllium oxide or other ceramic [7, 8], which can be water-cooled (e.g. Pharmacia Multiphor II) or an electrically cooled Peltier-effect cooling device (e.g. E-C

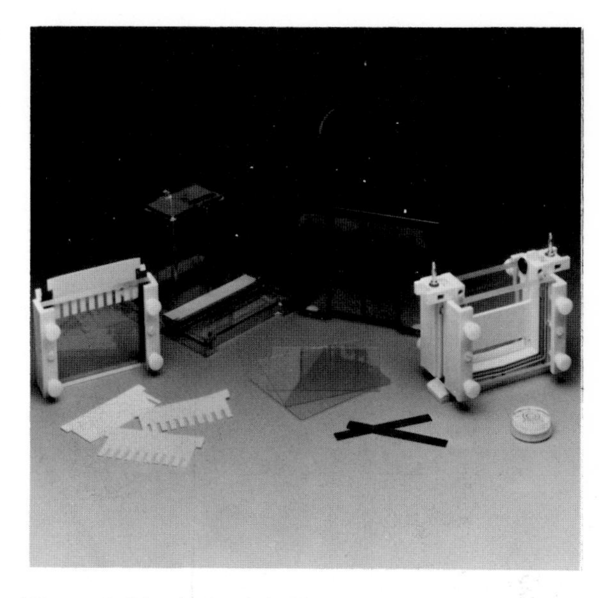

Figure 2. Vertical minigel apparatus

1001). For these cooling devices, heat transfer can be improved if the gel is cast on to a thin plastic backing sheet rather than on to a glass plate.

One significant problem, not encountered in vertical electrophoresis, is that gels tend to dry out during electrophoresis. The main causes of drying are evaporation (which is accelerated by heating) and electro-endosmosis (the field-induced movement of water and hydronium ions toward the cathode). Drying due to evaporation can be minimized by effective cooling. Drying due to electroendosmosis can be the result of the poor quality of agarose or inadequate wicking. The most common way to apply the electrical field is to place the electrodes into the buffer reservoir which is connected to the gel by paper wicks. A second common way is to use buffer-saturated filter papers, placed at the edges of the gel, as the electrolyte reservoirs themselves, with the electrodes placed directly on to the filter paper wicks. The capacity of paper wicks to move sufficient electrolyte into the gel can sometimes be exceeded by the rate of water loss. In order to cope with this, the use of 12%T, 3%C polyacrylamide or 1.5–3% (w/v) low endosmosis agarose 'cubes' saturated with electrolyte has been suggested [9]. These are placed directly on to the edges of the gel surface and can hold a large amount of liquid. A consequence of drying is that discontinuous gels, such as those employing stacking gels (see Introduction to Chapter IV and *Protocol 13b*), have not been routinely used (until recently) in the horizontal format. This is because the interface between the stacking and separating gels is especially vulnerable to drying; once dried, the seam opens and causes a current discontinuity. 'Seamless' gels are now prepared in which the two gels can have different buffer compositions (refs 10, 11 and *Protocol 13b*).

The second advantage of horizontal gels over vertical electrophoresis, is that sample loading is not restricted to one end of the gel. Since the sample can be loaded anywhere on the gel surface, buffer systems in which the proteins migrate in two directions can be used. This can be of particular value in isoelectric focusing experiments where loading the sample near its isoelectric point results in considerable time-saving.

When applications using continuous buffer systems (explained in Introduction to Chapter IV) are performed, it is possible to employ 'submarine' electrophoresis. In this procedure, since the buffer in the gel is the same as that in the electrolyte compartments, the horizontal gel can be submerged in electrolyte; mixing of discontinuous buffers is not an issue (*Figure 3*). Complete submersion of the gel is advantageous for even, efficient heat dissipation.

Where possible, the author advocates the use of rehydratable slab

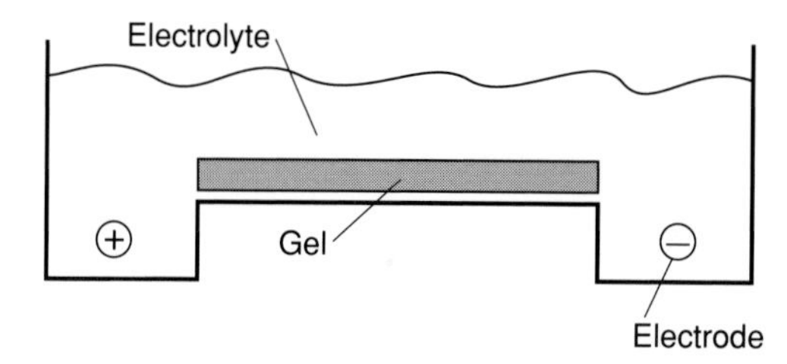

Figure 3. Submarine electrophoresis apparatus.

gels for isoelectric focusing and other horizontal electrophoresis applications where two-stage gels are not required [12,13]. Rehydratable gels have two important advantages over conventional gels. Firstly, they can be prepared in advance and stored dry for long periods. Secondly, impurities such as unpolymerized acrylamide, acrylamide breakdown products and unreacted initiators (ammonium persulfate, TEMED, riboflavin, etc.) are removed. The strategy for rehydratable gels is to cast the gels on a glass or plastic backing with just enough buffer salts to allow polymerization. Once gels are polymerized, the gels are soaked in water to remove buffers and impurities and then allowed to air dry. The dried gels can be stored at room temperature for long periods. Immediately before use, the gels are rehydrated by soaking in the appropriate electrolyte buffers.

Electrophoresis in cylindrical gels (see *Protocol 5c*)
Cylindrical gels (also called rod gels or tube gels) are the preferred geometry for applications which require a relatively long run time (overnight, for example), since lateral diffusion is limited by the walls of the cylinder. Such applications might be low-voltage separations or first-dimensional isoelectric focusing for two-dimensional experiments. Slab gels can permit appreciable lateral diffusion of the protein bands during lengthy runs because there is no force to contain the proteins in the dimension perpendicular to the electric field.

Gradient gels (see *Protocol 8*)

The rate at which a protein migrates through a semiporous matrix depends, in part, on the relationship of the size of the protein to the size of the pores. At one extreme, small proteins move virtually unrestricted through gels with very large pores and, at the other extreme, large proteins will be trapped in very small pores and cease migration. Between these extremes, one can manipulate the pore size of gels to improve the resolution of otherwise closely migrating proteins.

When only a few proteins are present in the sample, one can frequently tailor a single pore-size gel to achieve a good separation. However, when the sample contains many proteins, the use of a gradient gel, in which the pore size decreases from the top to the bottom may be more appropriate. Gradient gels offer two main advantages over single-concentration gels. Firstly, by shaping the gradient, one can simultaneously spread out or compress the pattern in different areas along the gel. Secondly, band sharpening occurs in gradient gels because the trailing edge of the band moves faster (through larger pores) than the leading edge.

A gradient gel is constructed by increasing the %T concentration from top to bottom. Most pore gradients are continuous; step gradients are rarely used because differently migrating proteins sometimes accumulate at the interfaces of the steps. The %T can vary linearly or exponentially. The gradients can be cast in the slab or cylindrical configuration, but the majority are in slabs.

Gradient gels can be cast using ordinary gradient makers, of the type usually used for preparing density gradients for centrifugation. Two significant points distinguish polyacrylamide gradient-making from gradient-making for centrifugation. Firstly, gradient makers work best when they can rely on differences in density between solutions of different solute concentration. The difference in density between a 5%T acrylamide solution and a 30%T acrylamide solution, for example, is not appreciable. Consequently, the more concentrated acrylamide solution is made denser by adding glycerol or sucrose. The second point is that the pouring of the gradient must be completed before the polymerization reaction begins. The most common gradient makers are two-chambered (*Figure 4*); for a discussion of multichambered gradient makers see ref. 14. The three basic designs for gradient makers are two open chambers, one open and one closed chamber, and the septum type. For all three designs the use of a peristaltic pump is recommended because it is generally more rapid than gravity feeding.

The simplest of the three designs (*Figure 4b*) can be made from ordinary laboratory materials and requires no machining of parts. This is the one having one open and one closed chamber. Only concave exponential gradients can be made with this device. Gradient makers

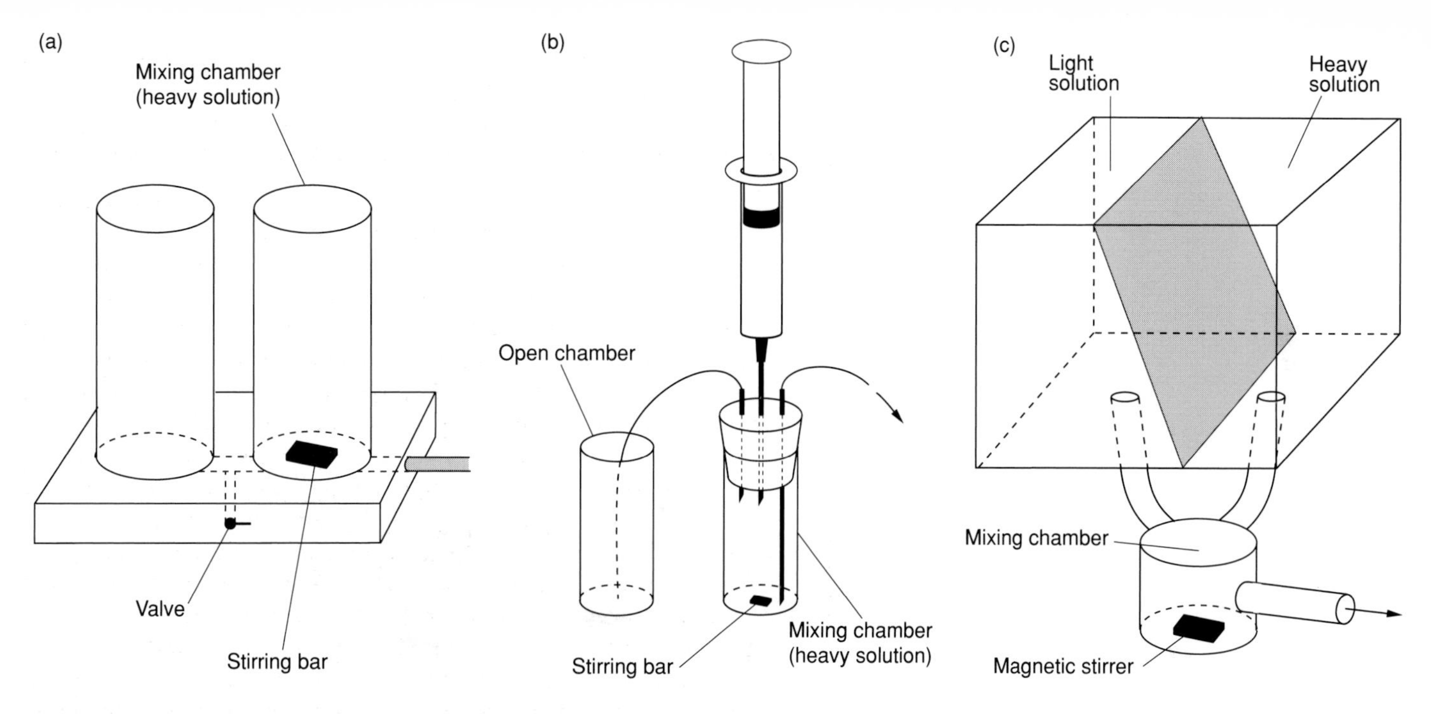

Figure 4. (a) Gradient maker with two open chambers. (b) Gradient maker with one open and one closed chamber for exponential gradients. (c) Gradient maker with septum-style gradient maker.

with two open chambers can be easily made with minimal machining of Plexiglass (also called Perspex or Lucite). They are relatively inexpensive when purchased ready-made. Linear gradients can be formed using two open chambers when both chambers are cylindrical. These designs rely on the fact that the two chambers are connected at the bottom. If the two chambers are cylindrical and have the same cross-sectional area, a linear gradient will result. A convex or concave exponential gradient can be prepared if one chamber is cylindrical and one conical. If the heavy solution is contained in the cylindrical mixing chamber, a convex gradient will result. A concave gradient will be formed if the heavy solution is contained in the conical mixing chamber.

It is possible to make perfectly linear gradients without the use of gradient makers [15] by the tilting method (*Protocol 8*) shown in *Figure 5*.

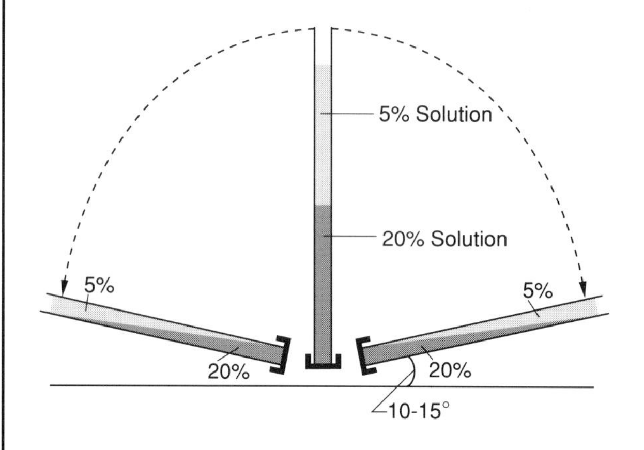

Figure 5. Tilting method.

Protocol 3. Preparation of glassware

Reagents

Acetone ⚠
18 M H_2SO_4 or Chromerge (Fisher Scientific) ⚠
No Chromix (Godax Labs)
Photoflow (Eastman Kodak)
RTV Silicone (General Electric or Dow)

Procedure

To clean glass plates

1 Degrease the plates by washing with acetone. ① ②

2 Wash the plates with soap and water.

3 Soak the plates overnight in an acid bath containing either 'Chromerge' or 18 M H_2SO_4 plus 40 g/l 'Nochromix'. [1]

4 Touching only the edges of the plates, rinse in tap water then in demineralized water.

5 Allow to drip dry and store wrapped in paper towel.

To clean glass tubes

Follow steps 1–4 as above.

Notes

This procedure will take approximately 16 hours.

① Plates which have been silanized (*Protocol 4*) can be re-used, but acetone washing is recommended for previously silanized plates.

② If silicone or other grease is used to render plates with spacers leak-proof, the plates should be degreased with acetone prior to washing.

③ Extrusion of cylindrical gels from glass tubes can be difficult if the glass is not well cleaned. To facilitate extrusion, many workers coat the tubes with a wetting agent. An alternative is to precoat the tubes with wax [16].

④ Many workers find it convenient to fix plastic spacers permanently on to the electrophoresis plates. The adhesive of choice is RTV silicone rubber (evolves acetic

5 Dry using reagent grade acetone, or

6 Dip tubes into a solution of 1% (v/v) 'Photoflow' (Kodak), which is a wetting agent, and allow to drip dry.③

To fix spacers permanently to glass plates

1 Dispense a bead of RTV Silicone sealer on to the plastic spacer and press the spacer and glass plate together.④

2 Allow to cure overnight.

3 With a razor, trim away any dried excess sealer which may have seeped on to the plate as this can sometimes interfere with acrylamide polymerization.

acid on setting) because the bond is resistant to repeated acid washings.

Pause point

[1] Plates can be left longer if convenient.

Protocol 4. Preparation of backing sheets

Reagents

Agarose
50% Ethanol ⚠
3-Methacryloxypropyl, trimethylsilane (also called Polifix 100, Serva Fine Biochemicals, or Bind Silane, Pharmacia) ⚠

Equipment

Boiling water bath or microwave oven
Electrophoresis plates
Silane plastic

Procedure

Polyacrylamide gel electrophoresis

1 To silanize glass plates, mix 2 ml of 3-methacryloxypropyl, trimethylsilane with 1 l of 50% ethanol, and allow to stand for 30 min.

2 Immerse the plates for 3–5 min, then allow to drip-dry vertically. Plates coated in this way can be stored at room temperature indefinitely. ① ②

Agarose electrophoresis

1 Prepare a suspension of 10 mg agarose per ml of distilled water.

2 Heat the suspension in a boiling water bath or microwave oven until the agarose is fully molten. Allow the agarose to cool to 10°C above its

Notes

This procedure will take approximately 30 min.

① When using silanized plates, make sure only one of the two plates used for casting the gel is silanized – otherwise you will never be able to separate them!

② The alternative to silanized glass is to use polyester plastic backing sheets. These can be purchased in large sizes, or precut to fit many sizes of commercial electrophoresis units, as GelBond-PAG (FMC Bioproducts) or Gel-Fix for PAGE (Serva). Plastic backing sheets for agarose are also available [GelBond (FMC Bioproducts) and Gel-Fix for Agarose (Serva)].

③ Before dispensing, it is necessary to ensure that the agarose is fully molten. Melting of the agarose suspension gelling temperature. See *Protocol 5d.* ③ ④

3 Dispense approximately 0.5 ml of molten agarose per 250 cm^2 of area on to the plate. Wipe the agarose over the plate with a lint-free tissue, and allow the surface to air-dry.

is indicated by the beaded shape disappearing.

④ Agarose, once melted and gelled, can be remelted several times. A stock for coating plates and for the electrophoresis matrix itself can be prepared in advance and stored in solidified form in a closed container. Be careful not to boil the stock because this will reduce the volume.

Protocol 5a. **Gel casting for vertical polyacrylamide slab gels**

Reagents

Choose recipes for gel solution from *Protocols 9–12*

Equipment

Casting stand ①②
Celloseal grease (Fisher Scientific)
Electrophoresis plates
Silanized plastic backing sheets
Spacers ③
Vacuum pump

Procedure

1 Check the electrophoresis plates for chipped edges which may cause the gel-forming cassette to leak. ④

2 If using a plastic silanized backing sheet, roll it on to one of the plates. To accomplish this, cut the plastic sheet to the precise size of the glass. Hold the edge of the sheet at 45° to the edge of the plate. If the plastic has both sides silanized (GelFix for PAGE), introduce a bead of water between the glass and the plastic. If the plastic has one side coated (GelBond), make sure the hydrophobic side is next to the glass and use a bead of 0.5% Triton X-100 instead of water. Lower the plastic sheet on to the glass and press the water to a film as the plastic is lowered as a flap on to the glass. Remove the excess water from between the glass and plastic using a rubber roller, or a test tube wrapped with a paper

Notes

This procedure will take approximately 1½ hours.

① This protocol is written for use with instruments having a separate casting stand (such as the BioRad Protean II).

② Slab gels to be run vertically are always cast vertically.

③ Analytical slab gels are typically 0.5–3.0 mm thick and use spacers usually made from polyvinyl chloride or other plastic. Natural or silicone rubber spacers make the cassette more easily watertight but deformability makes them less common.

④ When using silanized plates, coat only one. For systems using one notched and one rectangular plate, silanize the rectangular one. For systems with plates of unequal sizes, silanize the larger one. When using a silanized plastic backing sheet, roll it on to the rectangular plate, if one is notched, or the larger one if different sizes are used.

towel to roll the plastic surface. If the sheets do not contact the glass at all points, the gel will have uneven thickness when polymerized.④

3 Assemble the plates and spacers into a 'cassette', using a flat surface to align the bottom of the spacers with the bottom edge of the glass. It will leak if alignment is incorrect.⑤

4 Insert the cassette into the casting stand and tighten it against the rubber bumper using the off-center cams. Leak test the cassette by filling it with water.⑤ Do not remove the clips.⑥ Assembled apparatus can be left overnight.[1]

5 Vacuum de-gas the acrylamide solution, because gas bubbles will sometimes evolve during the polymerization (can be a problem for thin slabs). Connect the test tube or flask to a suction device. Maintain the suction until no more bubbles evolve. *Slowly* disconnect the vacuum.⑦

6 If a stacking gel is not to be used, pour or pipette the acrylamide solution, complete with buffers, initiators and accelerators, to the top of the cassette and insert a well-forming comb. If a stacking gel is to be used, fill the cassette high enough to allow for 1 cm of stacking gel beneath the teeth of the comb. Carefully pipette a 3 mm layer of distilled water or distilled water-saturated butanol on top of the acrylamide solution. Allow to polymerize.[2] See Introduction to Chapter IV for a discussion of stacking gels.⑧

7 When stacking gels are used, once polymerization of the separating gel is complete, pour off the water layer, fill the cassette to the top with solution for the stacking gel, and insert the well-forming comb. A well-forming comb which fits snugly into the cassette, will usually exclude enough air to permit polymerization all the way to the top of the cassette.

⑤ Most leakage occurs because the spacers are not completely flush with the edges of the glass at the bottom. If the cassette cannot be made leakproof by realignment, grease the spacers with a nonconductive grease (Celloseal). Do not allow grease on the inner face of the cassette. If leakage is due to chipped plates, grease the bottom edge of the plates. Alternatively, make a bead of molten 1% (w/v) agarose in water where glass meets bumper.

⑥ The clips holding the plates and spacers together cannot be removed until the electrophoresis is complete. Otherwise, the glass will flex, causing the gel and glass to separate. A short circuit can result.

⑦ When de-gassing, the acrylamide solution should contain buffers and TEMED, but not initiator. This takes a few minutes but can be left to stand. Immediately prior to introduction of the acrylamide solution into the cassette, add in the ingredient that has been omitted and mix gently.

⑧ The gel will polymerize completely only in the absence of oxygen. When discontinuous gels are used, the separating gel, which can be cast the day before use, is covered with a thin water layer. If longer than overnight elapses before casting the stacking gel, substitute the gel buffer (without mercaptoethanol) for the water layer.

Pause point

[1] Assembled apparatus can be left overnight.
[2] Can be left overnight.

Protocol 5b. **Gel casting for horizontal polyacrylamide slab gels**

Reagents

Choose recipes for gel solution from *Protocols 9–12*

Equipment

Binder or bulldog clips
Celloseal grease (Fisher Scientific)
Electrophoresis plates
Silanized plastic backing sheets
Spacers
Vacuum pump

Procedure

Vertical casting

1 Prepare a U-form spacer from plastic or rubber (see note 3 of *Protocol 5a*). A spacer at least 1 cm wide is recommended. ①

2 Assemble the plates with a U-form spacer such that the spacer is near the edge of the plates. ②③④

3 Clamp the plates and spacer together using binder clips or bulldog clips (as shown in *Figure 6*). These should completely surround the perimeter to prevent leakages.

Notes

This procedure will take approximately 1½ hours.

① For gels substantially thicker than 1 mm, horizontal casting is not recommended.

② When using plastic backing sheets, see step 2 of *Protocol 5a*. Glass which is much larger than the U-form will flex once the cassette is clamped, causing the plastic to buckle.

③ If silanized plates are used, coat only one. See note 1 of *Protocol 4*.

④ If leakage occurs, grease the U-form lightly with Celloseal on both sides.

⑤ To cast horizontally, both the sliding and flap techniques are used. The sliding technique is recommended for

4 Vacuum de-gas the acrylamide mixture (see *Protocol 5a*, step 5 and note 7).

5 Calculate the appropriate volume of mixture required for the gel, leaving 0.5 cm at the top for a water layer (see note 8 of *Protocol 5a*). Pour or pipette the acrylamide mixture into the U-form and then carefully pipette distilled water or distilled water-saturated butanol on top of the acrylamide solution to a depth of 3 mm. [1]

Horizontal casting ⑤

For the sliding technique.

1 Vacuum de-gas the acrylamide mixture (see *Protocol 5a*, step 5 and note 7).

2 Place the silanized or plastic-coated plate on a level surface. ⑥

3 Grease two spacers, which are the size of the long dimension, on one side each and place them at the edges of the plate with the greased side down. Press on the spacers to stick them to the plate.

4 Take the nonsilanized plate and hold it parallel to the table such that about 1 cm is resting on the spacers, and the rest is over the edge.

5 Pour the acrylamide solution gradually into the space where the plates overlap whilst, with a constant motion, sliding the upper plate along the spacers until the two plates are even. The ends are left open. Gels cast

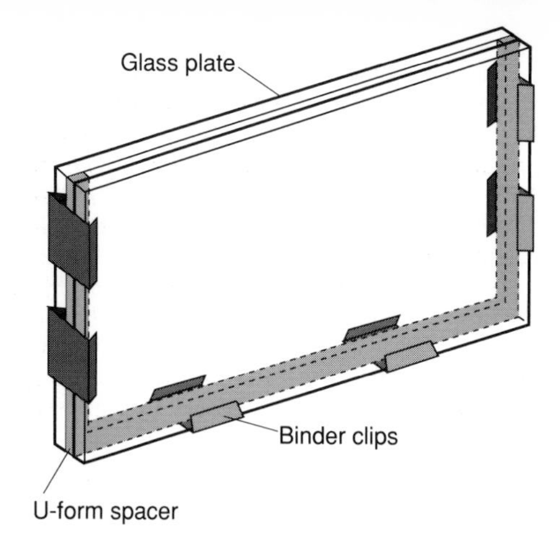

Figure 6. Vertical mold.

the majority of gels (0.2–1 mm thick), while the flap technique is recommended for ultrathin gels (0.06–0.2 mm thick) [17].

⑥ When using horizontal gels which are open on one face, it is advantageous to ensure that the gel does not stick to the plate to be removed. This can be done by: (i) silanizing one plate, (ii) using a silanized plastic backing sheet, (iii) substituting a nonwettable plastic plate for one of the glass by the sliding technique will polymerize to within a few millimeters of the ends. ⑥⑦

For the flap technique. ⑧

1 Vacuum de-gas the acrylamide mixture (see *Protocol 5a*, step 5 and note 7).

2 Place a thick glass plate with spacers on a flat surface (*Figure 7*). ⑨

3 Make a V by holding a second plate supporting a plastic backing sheet at a 45° angle to the horizontal.

4 Calculate the volume of liquid required for the gel and pour 150% of this volume on to the horizontal plate at the vertex of the V. The liquid will be held by surface tension.

5 Slowly (to avoid trapping air bubbles) lower the upper plate on to the spacers. Trapped air bubbles can usually be removed by repeatedly opening and closing the flap. Do not clamp the plates together; the glass can flex to a greater extent than the thickness of an ultrathin gel! ⑩ [2]

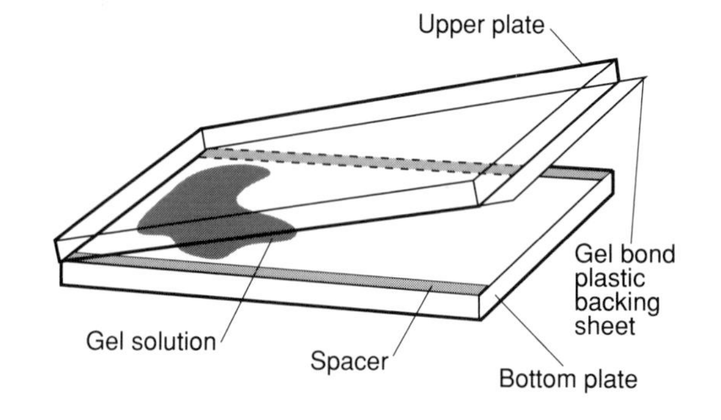

Figure 7. Flap technique.

plates, (iv) covering one plate with Gelbond film, putting the hydrophilic side to the glass and the hydrophobic side to the gel.

⑦ Gels cast by the sliding technique should be used the same day since the edges open to the air will dry out.

⑧ Ultrathin gels are cast by the flap technique: always cast on to backing sheets. Silanized glass is sometimes used but plastic is preferred. Both plates should be thick glass, which is usually flatter than thin glass.

⑨ For spacers for ultrathin gels use Parafilm (single thickness is approx. 120 μm). The use of polythene sandwich bags of approx. 60 μm has also been reported.

Pause points

1. Gels may be cast the day before use (see note 8 of *Protocol 5a*).
2. Ultrathin gels cast by the flap technique are sometimes used immediately. Frequently they are used as rehydratable gels (see *Protocol 6*).

Protocol 5c. **Casting cylindrical polyacrylamide gels**

Reagents

Choose recipes for gel solution from *Protocols 9–12*

Equipment

Gel tubes (see *Protocol 3*)
Parafilm
Syringe and needle
Test tube rack

Procedure

1. Cover the bottom of each prepared glass tube (*Protocol 3*) with Parafilm to make a water-tight seal and place vertically in a rack. ①
2. Follow steps 5–7 in *Protocol 5a*, with the following modification. In order to avoid air locks in the cylinder, introduce the acrylamide solution into the tube using a syringe fitted with a large-bore needle (at least 16 gauge). If the needle does not reach the bottom of the tube, attach a plastic tube to the end of the needle. Be careful to introduce the acrylamide solution slowly so as not to create additional bubbles. The stacking gel will polymerize to within a few millimeters of the top.

Notes

This procedure will take approximately 1½ hours.

① Use tubes with heat-smoothed ends. Gentle warming of the tube will facilitate sealing.

Protocol 5d. **Agarose slab gel electrophoresis**

Reagents

Buffer solution (choose recipes from *Protocols 9–12)*
Powdered agarose

Equipment

Boiling water bath or microwave oven
Electrophoresis plates
Leveling table
Waterproof tape

Procedure

1 If plastic backing sheets are used, roll them on to the plates as directed in step 2 of *Protocol 5a*. If not, precoat the glass plate with agarose as directed in *Protocol 4*.

2 Weigh out the appropriate amount of agarose, add buffer solution and melt the agarose in a hot water bath or microwave oven, at the temperature suggested for that particular agarose. Agaroses come in a variety of different types, which melt and solidify at different temperatures; the melting temperature is usually considerably higher than the gelling temperature (refs 4,5, Chapter I). See notes to *Protocol 4.* ①

3 Place the plate on a leveling table, pour or pipette the agarose solution on to the center of the plate after it has cooled to within 10–15°C of its gelling temperature. If necessary, spread the agarose to the edges with the end of a pipette. ② [1]

Notes

This procedure will take approximately 1 hour.

① If heat-sensitive ingredients are to be included in the gel, do not add them until the molten agarose has cooled to 10–15°C above its gelling temperature.

② Agarose gels are rarely cast between two parallel plates. For gels of 1 mm thickness or less, surface tension is sufficient to hold the molten agarose in place until it solidifies. Agarose gels thicker than about 1 mm require a wall around the plate to contain the agarose until it solidifies. Prepare such a wall by applying waterproof masking tape around the perimeter of the plate.

Pause point

[1] Agarose plates may be prepared in advance of use, but they must be maintained in a humidified chamber until use. A closed plastic food storage container, containing water-saturated paper towels, makes a convenient humidor.

Protocol 6. Rehydratable gels

Reagents

Acrylamide/AcrylAide stock – 30 g acrylamide ⚠, 40 ml of 2% (w/v) AcrylAide solution (FMC Bioproducts) made up to 100 ml with distilled water ①
5% (w/v) Ammonium persulfate in distilled water
TEMED ⚠
0.5 M Tris-HCl, pH 8.0

Equipment

Binder or bulldog clips
Electrophoresis plates
Shallow dish or rehydration cassette ②
Silanized plastic backing sheet
U-form spacer

Procedure

1 This gel is cast in the vertical direction using a U-form spacer. Assemble the plates and backing sheet according to *Protocol 5b*.

2 To make 5 ml of gel, mix :
- 0.25 ml of acrylamide/AcrylAide solution
- 0.5 ml 0.5 M Tris-HCl, pH 8.0
- 50 μl ammonium persulfate solution
- 4.25 ml distilled water.

3 Vacuum de-gas (*Protocol 5a*, step 5 and note 7) then add 0.05 ml TEMED and mix gently. Pour the solution between the plates and overlay with a distilled water layer.

Notes

This procedure will take approximately 2 hours, plus overnight drying.

① Rehydratable gels dried on plastic sheets curl less when AcrylAide is the cross linker instead of bisacrylamide. Note that proteins migrate faster in gels cross linked with AcrylAide; an AcrylAide gel which is 12%T, 2.66%C is approximately equivalent to a 10%T, 2.66%C bisacrylamide gel.

② When rehydrating use at least 120% of the final volume required by the gel. The extent of the rehydration varies with time and %T. If excessive, the gel will swell and can reach as much as 150% of its original thickness [18]. To

4 Once polymerization is complete, soak the gel and its backing in a dish of water. For gels bonded to plastic, soak face-down for 20–30 min, depending on the thickness. A 0.5 mm gel requires about 20 min. Alternatively, if the gel is cast on to a glass plate, soak it face-up.

5 Allow the gel to air dry. [1]

6 Prior to use, rehydrate the gel by soaking in the appropriate electrolyte or ampholyte solution. This is done either in a dish slightly larger in area than the gel ② or in a rehydration cassette made from two glass plates and a rubber U-form spacer. ③ [2]

determine when to terminate the rehydration, weigh the gel before drying and then periodically during the rehydration. Rehydration strategies are discussed in refs 18 and 19.

③ Cassettes ensure that the rehydrated gel does not exceed its original thickness. Cut a spacer whose thickness equals that of the original spacer plus backing sheet. The distance between the verticals of the U should be slightly wider than the width of the gel so as to be economical of rehydration solution. Clamp plates and spacer together, insert the dried gel into the U and add rehydration solution.

Pause points

[1] The paper-thin dried gel can be stored for 6 months in a dust-free environment at room temperature.

[2] Rehydration in a cassette can be left overnight.

Protocol 7. **Preparation of gradient gels using gradient makers**

Reagents

Acrylamide/bisacrylamide stock – 30g acrylamide ⚠, 0.8 g bisacrylamide ⚠ made up to 100 ml with distilled water
5% (w/v) Ammonium persulfate in distilled water
Glycerol
5× Separating gel buffer solution. For Tris-glycine gels, use the 5× solution given in *Protocol 9*. For SDS gels, use the solution in *Protocol 12*
5× Stacking gel buffer, if required. For Tris-glycine gels, use the 5× solution given in *Protocol 9*. For SDS gels, use the solution in *Protocol 12*
TEMED ⚠

Equipment

Electrophoresis plates
Gradient maker (e.g. Isco, Pharmacia or homemade)
Magnetic stirrer and stirring bar
Peristaltic pump (e.g. Buchler or Pharmacia)
Silanized plastic backing sheets
Spacers

Procedure

To cast a 14 ml 5–20%T, 2.66%C exponential gradient using a gradient maker with one open and one closed chamber ①②⑥

1 Assemble the gradient maker shown in *Figure 4b* or construct one from a disposable syringe [21]. Alternatively, a gradient maker having two open chambers can be converted to one having one open and one closed

Notes

This procedure will take approximately 1½ hours.

① For casting two or more gradients simultaneously, a 'stream splitter' should be inserted in the outlet line.

② The appropriate volume for many standard electrophoresis cells is 14 ml (e.g. BioRad Protean II with a 0.75 mm spacer).

chamber by inserting a vented rubber stopper into the top of the mixing chamber, then closing the vent.

2 Assemble the plates, spacers and backing sheets, if used.

3 Prepare 7 ml of the heavy solution by mixing:
- 4.7 ml of acrylamide/bisacrylamide stock solution
- 1.4 ml of 5× separating gel buffer solution
- 0.2 ml distilled water
- 0.7 ml glycerol. ③

4 Prepare 14 ml of the light solution by mixing:
- 2.4 ml of acrylamide/bisacrylamide stock solution
- 2.8 ml of 5× separating gel buffer solution
- 8.8 ml distilled water.

5 Vacuum de-gas (*Protocol 5a*, step 5 and note 7) the heavy and light solutions and cool them down to 4°C.

6 Add 50 μl of ammonium persulfate solution to the heavy solution and 100 μl to the light solution. Swirl gently to mix. If TEMED is not included in the 5× buffer solution, it should be added now. ④

7 Put a magnetic stirring bar into the mixing (closed) chamber and pour in the heavy solution. Close the chamber, put the gradient maker on to a magnetic stirrer and stir. ⑤⑥

8 Put the light solution into the open chamber. If using the gradient maker

③ To assess the gradient's shape (linear, sufficiently convex, etc.), add about 5 mg/ml Blue Dextran (Pharmacia) (which will not diffuse following polymerization) to the heavy solution before casting. After polymerization, measure the gel color with a densitometer or scanning spectrophotometer.

④ These gel recipes have TEMED included in the separating gel buffers. This is preferable since accurate pH adjustment is facilitated. For recipes which do not premix TEMED, add TEMED at the times indicated.

⑤ Gradients will be reproducible only if the mixing is complete. Stir as fast as is practicable without introducing air bubbles.

⑥ See ref. 20 for additional recipes for exponential gradients.

in *Figure 4b*, withdraw the plunger of the syringe slowly to fill the tubing connecting the chambers with light solution. If using a gradient maker with a valve between the chambers, open it now.

9 Hold the outlet tube of the peristaltic pump at the top of the gel cassette and begin pumping slowly (about 2 ml/min). Discontinue pumping when the open chamber is empty.

10 Add a water layer to the top of the gradient (see note 8 of *Protocol 5a*). [1]

11 Clean the gradient maker by filling both chambers with water and pumping the system through *before polymerization begins*.

12 When stacking gels are used, once polymerization of the separating gel is complete (an interface will appear), pour off the water layer, fill the cassette to the top with solution for the stacking gel, and insert the well-forming comb. Prepare the 3%T, 2.66%C stacking gel solution by mixing, immediately prior to use:
- 1.0 ml acrylamide/bisacrylamide solution
- 2.0 ml 5× stacking gel buffer
- 50 μl ammonium persulfate solution
- 50 μl TEMED.

To cast a 14 ml linear 5–20%T, 2.66%C gradient, using a gradient maker with two open chambers (1) (2)

1 Assemble the plates, spacers and backing sheets, if used.

2 Prepare 7 ml of the heavy solution according to step 3 previously described. (3)

3 Prepare 7 ml of the light solution by halving the amounts given in the recipe in step 4 previously described.

4 Vacuum de-gas (*Protocol 5a*, step 5 and note 7) the heavy and light solutions. To ensure that pouring is complete before the onset of polymerization, cool the solutions down to 4°C.

5 Add 50 μl of ammonium persulfate solution to the heavy solution and 50 μl to the light solution. Swirl gently to mix. If TEMED is not included in the 5× buffer solution, it should be added now.④

6 Close the valve connecting the chambers and clamp off the exit port to the mixing chamber (the chamber connected to the peristaltic pump).

7 Put the gradient maker on to a magnetic stirrer, place a stirring bar into the mixing chamber and pour in the heavy solution. Put the light solution into the other chamber.

8 Switch on the stirrer and start the solution stirring in the mixing chamber.⑤ Open the valve, open the clamp and perform steps 9–12 above.

Pause point

1 The storage and use conditions for gradient gels are the same as those for single concentration polyacrylamide gels.

Protocol 8. Preparation of gradients by the tilting method

Reagents

Acrylamide/bisacrylamide stock – 30 g acrylamide ⚠ , 0.8 g bisacrylamide ⚠ made up to 100 ml with distilled water
5% (w/v) Ammonium persulfate in distilled water
Glycerol
5× Separating gel buffer solution. For Tris-glycine gels, use the 5× solution given in *Protocol 9*. For SDS gels, use the solution in *Protocol 12*
5× Stacking gel buffer, if required. For Tris-glycine gels, use the 5× solution given in *Protocol 9*. For SDS gels, use the solution in *Protocol 12*
TEMED ⚠

Equipment

Casting stand ①
Electrophoresis plates
Silanized plastic backing sheets

Procedure

1 To prepare a linear 5–20%T, 2.66%C gradient by the tilting method, assemble the plates, spacers and backing sheets, if used. ②

2 Prepare 28 ml of the heavy solution by mixing:
- 18.8 ml of acrylamide/bisacrylamide stock solution
- 5.6 ml of 5× separating gel buffer solution
- 0.8 ml distilled water
- 2.8 ml glycerol.

3 Prepare 28 ml of the light solution by mixing:
- 4.8 ml of acrylamide/bisacrylamide stock solution

Notes

This procedure will take approximately 1½ hours.

① This procedure is best performed when gels are prepared on a casting stand.

② The appropriate volume for many standard electrophoresis cells is 56 ml (e.g. BioRad Protean II using a 3 mm spacer).

③ Since forming a gradient by tilting takes slightly longer than pumping one, ensure that the solutions and plates are fully cooled.

④ It is necessary to leave at least 3 cm at the top of the cassette to allow tilting without spilling.

- 5.6 ml of 5× separating gel buffer solution
- 17.6 ml distilled water.

4 Vacuum de-gas (*Protocol 5a*, step 5 and note 7) the heavy and light solutions and cool them down to 4°C. Precool the gel cassette once it has been assembled. This is to retard the onset of polymerization.

5 Add 200 μl of ammonium persulfate solution to the heavy solution and 200 μl to the light solution. Swirl gently to mix. If TEMED is not included in the 5× buffer solution, it should be added now (see note 4 to *Protocol 7*).③

6 Pour or pipette the heavy solution into the cassette. Carefully layer the light solution on top of the heavy solution.

7 Taking care to avoid spilling the solution, tilt the cassette toward the table until it forms an angle of about 10–15° with the table. Leave the cassette for 10 sec and then place the cassette upright for 10 sec. Tilt the cassette in the opposite direction for 10 sec and place the cassette upright again (see *Figure 5*, p. 32). Repeat this cycle 20 times.④⑤

8 Perform steps 10 and 12 of the exponential procedure of *Protocol 7*.

⑤ Do not be alarmed if the Schlieren pattern looks irregular during the tilting process. When righted, the gradient will orientate perfectly!

⑥ Gradients made by this method must be at least 2 mm thick.

IV GEL TECHNIQUES

Introduction

Separation of proteins based upon differences in their electrophoretic mobility in gels can be achieved in two fundamentally different ways. One may use the intrinsic charge of the proteins as the determinant of their electrophoretic mobility, or one may impart an extrinsic charge to the proteins. The latter case is almost always performed to achieve separation based on molecular weight. A greater variety of techniques is available for charge-based separations, which are generally grouped into three categories: continuous buffer systems, discontinuous buffer systems and isoelectric systems. Many variations exist within each category.

Methods available

Separations under native conditions using intrinsic charge

The intrinsic charge that a protein carries is determined by its amino acid composition and three-dimensional configuration. When two proteins vary widely in their acidic and basic amino acid compositions, a pH can usually be found at which their intrinsic charges are appreciably different. Thus separation can be readily achieved using a simple, continuous buffer system, where the buffers in the separating gel, the sample and the tank are all the same. In these systems, the gel is a one-stage gel and the sample is loaded directly on to it. An additional requirement for continuous buffer systems is that the sample be applied to the gel in a small volume. For a description of con-

References

1. McLellan, T. (1982) *Analyt. Biochem.* **126**, 94–99.
2. Chrambach, A. (1985) *The Practice of Quantitative Gel Electrophoresis*, pp. 7–11. VCH, Weinheim.
3. Ornstein, L. (1964) *Ann. N.Y. Acad. Sci.* **121**, 321–349.
4. Davis, B.J. (1964) *Ann. N.Y. Acad. Sci.* **121**, 404–427.
5. Deyl, Z. (1979) in *Journal of Chromatography Library*, vol. 18, pp. 119–121. Elsevier, Amsterdam.
6. Hedrick, J.L. and Smith, A.J. (1968) *Arch. Biochem. Biophys.* **126**, 155–164.
7. Goldenberg, D.P. and Creighton, T.E. (1984) *Analyt. Biochem.* **138**, 1–51.
8. Murch, R.S. and Budowle, B. (1986) *J. Forensic Sci.* **31**, 869–880.

tinuous buffer systems see ref. 1.

The majority of separations cannot be performed in such simple systems, either because the sample contains components that are closely related, because the sample contains many components, or because the samples are too dilute. In these instances, discontinuous buffer systems are required, along with the use of stacking gels. Proteins in solution are dispersed throughout the entire volume of solution. In order to achieve a good separation, it is necessary for the entire population of individual molecules of a single component of the mixture to enter the separating gel as a concentrated sample. Accordingly, if samples are not highly concentrated in a very small volume, a gel system must be used which will sort out the various components and order them in concentrated zones. Such systems are termed discontinuous buffer systems. They generally employ two-stage, abutting gels, each having a different pH. Different ions may also be contained in the tank buffer. For a detailed description of the theory of discontinuous buffer systems see ref. 2.

The upper, or stacking gel, where the sample is loaded, has fairly large pores (usually 3%T) so as to have a minimal effect on protein migration. Here the components of the mixture are concentrated and ordered in to a tight 'stack' using a buffer system where the net mobilities of the proteins are between those of the leading and trailing ion. The sieving effect of the lower (also called separating or resolving)

9. Prestidge, R.L. and Hearn, M.T.W. (1979) *Analyt. Biochem.* **97**, 95–102.
10. Cuono, C.B. and Chapo, G.A. (1982) *Electrophoresis*, **3**, 65–75.
11. Righetti, P.G. and Gianazza, E. (1992) *Electrophoresis*, **13**, 185–186.
12. Allen, R.C., Budowle, B., Lack, P.M. and Graves, G. (1986) in *Electrophoresis '86* (M.J. Dunn, ed.), pp. 462–473. VCH, Weinheim.
13. Radola, B.J. (1987) in *New Directions in Electrophoresis Methods* (J.W. Jorgenson and M. Phillips, eds), vol. 335. ACS Symposium Series.
14. Chiari, M. and Righetti, P.G. (1992) *Electrophoresis*, **13**, 187–191.
15. Fawcett, J.S. and Chrambach, A. (1988) *Electrophoresis*, **9**, 463–469.
16. Dossi, G., Celentano, F., Gianazza, E. and Righetti, P.G. (1983) *J. Biochem. Biophys. Meth.* **7**, 123–142.
17. Sinha, P., Kottgen, E., Westermeier, R. and Righetti, P.G. (1992) *Electrophoresis*, **13**, 210–214.
18. Laemmli, U.K. (1970) *Nature*, **227**, 680–685.
19. Shapiro, A.L., Vinuela, E. and Maizel, J.V. (1967) *Biochem. Biophys. Res. Comm.* **28**, 815–820.
20. Raines G., Aumann, H., Sykes, S. and Street, A. (1990) *Thromb. Res.* **60**, 201–212.
21. O'Farrell, P.H. (1975) *J. Biol. Chem.* **250**, 4007–4021.
22. Gorg A., Postel, W. and Gunther, S. (1988) *Electrophoresis*, **9**, 531–546.

gel allows the separation of 'stacked' proteins with different mobilities.

Tris-glycine gels (see *Protocol 9*)

Tris-glycine gels, according to the original formulation of Ornstein [3] and Davis [4] , are the most commonly used discontinuous, nondenaturing gels. The majority of proteins can be separated in this system. For very basic proteins, a low pH system is used [5].

Ferguson analysis

An alternative approach to molecular weight determination in SDS gels, uses proteins in native configuration. This is Ferguson analysis, in which the relative mobility of proteins is measured in polyacrylamide gels having different pore sizes (or in a single slab gel with a transverse gradient of pore sizes) [6, 7].

Problems: Although this approach removes the problem of anomalous binding of SDS, shape and glycosylation issues remain (see p. 60).

Electrofocusing

Many variations for electrofocusing exist including native and denaturing applications. Focusing under denaturing conditions is used most frequently as part of two-dimensional separations. Electrofocusing under native conditions is more common.

23. Cleveland D.W., Fischer, S.G., Kirschner, M.W. and Laemmli, U.K. (1977) *J. Biol. Chem.* **252**, 1102–1106.
24. Laine, A., Ducourouble, M.P. and Hannothiaux, M.H. (1978) *Analyt. Biochem.* **161**, 39–44.
25. Daerr, W.H., Minzlaff, U. and Greten, H. (1986) *Biochim. Biophys. Acta*, **879**, 134–139.
26. Lebecq, J.C., Salhi, S.L. and Bastide, J.M. (1984) *J. Immunol. Methods*, **66**, 219–226.
27. Allen, R.C. and Lack, P.M. (1987) in *New Directions in Electrophoretic Methods* (J.W. Jorgenson and M. Phillips, eds), pp. 117–131. ACS Symposium Series, Washington, DC.
28. Radola, B.J. (1983) in *Electrophoretic Techniques* (C.F. Simpson and M. Whittaker, eds), pp. 101–118. Academic Press, London.
29. Righetti, P.G. (1983) *Isoelectric Focussing: Theory, Methodology and Applications*, pp. 195–196. Elsevier, Amsterdam.
30. Strahler, J.R., Hanash, S.M., Somerlot, L., Weser, J., Postel, W. and Gorg, A. (1987) *Electrophoresis*, **8**, 165–173.
31. Gersten D.M., MacGregor, C.H., McElhaney, G.E. and Ledley, R.S. (1982) *Electrophoresis*, **3**, 231–232.
32. Nguyen, N.Y. and Chrambach, A. (1977) *Analyt. Biochem.* **82**, 226–235.
33. Kinzkofer-Peresch, A., Petestos, N.P., Fauth, M., Kogel, F., Zok, R. and Radola, B.J. (1988) *Electrophoresis*, **9**, 497.
34. Gorg, A., Postel, W., Baumer, M. and Weiss, W. (1992) *Electrophoresis*, **13**, 192–203.

As the net charge of a protein depends in part on the environmental pH, it follows that, for any given protein, there is a pH, at the isoelectric point (pI), where the net charge of the protein is zero, that is, it will not migrate in an electric field. Isoelectric focusing, therefore, seeks to create a pH gradient across the gel, in which the proteins will cease migration at the position in the gel where the pH corresponds to their pI. As an equilibrium procedure, isoelectric focusing is most appropriately performed in a large-pore gel in which molecular sieving is minimal (e.g. low %T polyacrylamide or agarose). Focusing to equilibrium can have a concentrating effect, and the potential to separate proteins which vary only slightly in amino acid composition or post-translational modification is considerable.

The pH gradients can be created using soluble molecules (buffer salts or carrier ampholytes), immobilized molecules (Immobilines) and mixtures of carrier ampholytes and Immobilines. Buffer electrofocusing, is described in detail elsewhere [9–11] .

Focusing with carrier ampholytes in tube gels

The most frequent isoelectric focusing experiments use gradients formed by carrier ampholytes, which are mixtures of amphoteric compounds. The strategy for separations using carrier ampholytes is first to cast a gel containing the ampholytes. Next the ends of the gel are connected to electrolyte reservoirs containing an anolyte (+) of

35. Anderson, N.G. and Anderson, N.L. (1978) *Analyt. Biochem.* **85**, 351–354.
36. Gersten, D.M., Ramagli, L.S., Johnston, D.A. and Rodriguez, L.V. (1992) *Electrophoresis*, **13**, 87–92.
37. Moos, M., Nguyen, N.Y. and Liu, T.Y. (1988) *J. Biol. Chem.* **263**, 6005–6008.
38. Gorg, A., Postel, W., Weser, J., Gunther, S., Strahler, J.R., Hanash, S.M. and Somerlot, L. (1987) *Electrophoresis*, **8**, 122–124.
39. Litteria, M. (1989) *Appl. Theor. Elec.* **1**, 265–266.
40. Glazer, A.N., Delange, R.D. and Sigman, D.S. (1976) *Chemical Modification of Proteins*, pp. 73–74. North-Holland, Amsterdam.

Protocols provided

- **9.** *Tris-glycine polyacrylamide gels*
- **10a.** *Isoelectric focusing in rehydratable polyacrylamide slab gels*
- **10b.** *Isoelectric focusing in agarose gels*
- **10c.** *Isoelectric focusing in ultrathin gels (pH 5–7)*
- **11.** *Isoelectric focusing using Immobilines*
- **12.** *SDS–polyacrylamide gels*
- **13a.** *Two-dimensional electrophoresis with carrier ampholytes using a cylindrical first dimensional gel and a vertical second-dimensional slab gel*
- **13b.** *Two-dimensional electrophoresis using horizontal Immobiline slabs in the first dimension and horizontal*

dilute acid and a catholyte (−) of dilute base. On application of the current, the ampholytes will form a pH gradient which is stable for several hours. Establishment of the pH gradient by the ampholytes is called prefocusing, following which the proteins are loaded. Focusing proceeds until protein migration ceases.

Problems: Isoelectric focusing experiments were originally performed exclusively in cylindrical gels in which the carrier ampholytes were mixed in with the acrylamide and bisacrylamide monomers prior to polymerization. A significant problem, is gradient drift, which is the major problem in terms of experiment-to-experiment comparison of results. As a consequence of long periods of focusing, the basic end of the gradient begins to migrate off the end of the gel. Gradient drift is caused primarily by electroendosmosis, the field-induced movement of water and hydronium ions toward the cathode. Charged groups on the walls of the glass tube are the major cause of electroendosmosis.

Focusing in horizontal slabs

Electroendosmosis is substantially reduced (but not completely eliminated) by the use of horizontal slab gels. A second advantage of the horizontal format is that rehydratable gels can be used (see *Protocol 10a*). This procedure removes unwanted substances (*Protocol 6*), which are more troublesome in isoelectric focusing than in other gel techniques. It is particularly useful for zymography experiments [12].

SDS slabs in the second dimension

14. *Peptide mapping by the Cleveland technique*

15. *Immunoelectrophoresis*

16a. *Rocket immunoelectrophoresis*

16b. *Crossed immunoelectrophoresis*

The use of horizontal ultrathin gels (60–120 μm thick – see *Protocol 10c*) has several advantages. Firstly, the amount of time required for the focusing is greatly reduced. This is because heat dissipation by very thin gels is efficient and so relatively high voltages can be applied across the gel. Separations which would take hours using conventional gels can be achieved in minutes [13]. Secondly, post-electrophoresis, fixing, staining and destaining times are reduced. The third advantage, is that, with diffusional factors kept to a minimum as a result of very short elapsed times, the bands are more highly concentrated, resolution is improved and so a better detection sensitivity is achieved even though less protein is loaded.

Focusing with Immobilines (see *Protocol 11*): One of the most significant improvements in electrofocusing has been the advent of immobilized pH gradients. In this procedure, the pH gradient is not formed in the electric field. Instead, it is cast into the gel. Tertiary amino and weak carboxyl groups are conjugated to acrylamide monomer. These conjugates, termed Immobilines, are then added to acrylamide monomer and bisacrylamide; a pH gradient, formed by a gradient-maker (see *Protocols 7* and *8* for gradient-making) is polymerized on addition of persulfate and TEMED. The structure of Immobilines has been described [14].

The use of a solid-phase pH gradient has the advantage of infinite (within reason) stability and the potential for high experiment-to-experiment reproducibility. Thus, focusing can be performed for extended periods of time without untoward effects. A further advantage is the ability to cast very narrow-range (within 0.1 pH unit) gradients. ***Problems***: Early experience with Immobilines was such that their use was somewhat more cumbersome than carrier ampholytes. One problem was that conductivity discontinuities over some pH ranges were troublesome. These could be corrected by adding carrier ampholytes to the system [15]. A second troublesome factor was that wide range gradients (spanning several pH units) had to be cast using an elaborate five-chambered gradient-maker [16]. However, procedures are now available for preparing a large variety of Immobiline gradients, without carrier ampholyte supplementation, using a simple two-chambered gradient maker [17]. There are two main disadvantages of focusing in immobilized pH gradients. First, although the procedures for preparing the Immobiline gradients have been simplified, they are still laborious. Second, focusing in immobilized pH gradients is more costly than using carrier ampholytes.

Separations under denaturing conditions

SDS gel electrophoresis

In this procedure, the sample is prepared and electrophoresed in the presence of SDS (full name sodium dodecyl sulfate, also called sodium lauryl sulfate) -containing buffers [18]. SDS is a negatively charged detergent which binds by its hydrophobic domain to hydrophobic amino acids. Each gram of the average protein binds 1.4 g of SDS, essentially swamping out the protein's intrinsic charge. Thus, the proteins attain a uniform charge per unit length. In an unrestricted electric field, such protein–SDS complexes would migrate at uniform velocity. However, when electrophoresed through a semiporous gel, the migration of the larger molecules is retarded relative to that of the smaller ones. A separation based on size is therefore achieved.

SDS gel electrophoresis according to the Laemmli technique is also possible in agarose gels [20] but it is almost always performed in polyacrylamide gels. For gels of the appropriate porosity, a plot of $\log_{10}$ molecular weight against R_f has a broad linear range [19]. R_f is the distance the protein migrates divided by the distance migrated by the leading ion. In this case the position of the leading ion can be located by the position of the Bromophenol Blue tracking dye front. Thus the molecular weight of an unknown protein is determined by measuring the mobility relative to those of a set of proteins of known molecular weight, and is designated M_r.
Problems: The determination of molecular weight by electrophoresis in SDS gels has its limitations, as do all methods short of amino acid sequence determination. Firstly, for precise M_r determination, all proteins would have to bind precisely the same amount of SDS per

unit size; proteins with atypical proportions of hydrophobic: hydrophilic amino acids will migrate anomalously. Secondly, all proteins must be the same shape. The inclusion of sulfhydryl-reducing agents such as 2-mercaptoethanol or dithiothreitol to reduce inter- and intra-chain disulfide bridges will minimize but not eliminate differences in shape. Thirdly, histones, glycosylated proteins, lipoproteins and other proteins having post-translational modifications may also migrate anomalously.

Two-dimensional electrophoresis

Two-dimensional electrophoresis, combining denaturing electrofocusing in the first dimension and pore gradient SDS electrophoresis in the second is among the most powerful tools of modern biology. The resolving power of the best one-dimensional gel techniques is in the range of 100 bands. Given that cells have been estimated to express as many as 10 000 proteins, analysis of mixtures as complex as that requires approaches with greater separating potential. Analysis of proteins and peptides in which the molecules are separated at right angles on the basis of two unrelated parameters has been known for many years but was popularized following the work of O'Farrell in 1975 [21]. Over the next few years it became clear that many artifacts can be introduced into the patterns by experimental techniques [22] and that reliably congruent two-dimensional patterns which allow straightforward comparison among gels can be very difficult to obtain. However, two-dimensional electrophoresis has evolved into a very useful technology.

Cylindrical first dimensions and vertical second dimensions (see *Protocol 13a*): Electrophoresis using carrier ampholytes in a cylindrical first dimension and a vertical second-dimensional slab is the most common of the two-dimensional techniques. Even considering the drawbacks of carrier ampholytes in glass tubes, many users will find this arrangement to be satisfactory for the majority of applications. The strategy for this procedure is to detergent-solubilize the proteins maximally, reduce them to their subunit structure with a sulfhydryl agent (mercaptoethanol or dithiothreitol), denature them using a polar solvent (urea), and disperse them using a nonionic organic detergent (NP-40). These conditions are maintained throughout first-dimensional electrofocusing. Following the focusing step, the first-dimensional gel is removed from its glass tube and equilibrated with SDS-containing buffers. The gel cylinder is laid on top of a pore gradient SDS slab gel for separation according to size.

Proteins with very basic pIs are sometimes separated by nonequilibrium focusing in the first dimension. In this procedure, the proteins are separated in a carrier ampholyte-containing gel with the anolyte and catholyte reversed. Both equilibrium and nonequilibrium first dimensions follow the same basic strategy. Adaptations required for nonequilibrium first dimensions are given in *Protocol 13a,* note 7.

Problems: The shortcomings are mainly associated with the use of a cylindrical first-dimensional gel containing carrier ampholytes. Firstly and most importantly, gel-to-gel and laboratory-to-laboratory reproducibility is difficult in view of gradient drift considerations and manufacturing differences between batches of carrier ampholytes. Secondly, the use of cylindrical polyacrylamide focusing gels with large pores and hence poor physical strength is difficult because of the need to transfer them to the second-dimensional gel. The use of smaller-pored gels increases dimensional stability but also increases focusing time, which in turn, increases gradient drift. Thirdly, electro-endosmosis occurs prominently in cylindrical gels and contributes substantially to gradient drift.

Horizontal two-dimensional electrophoresis (see *Protocol 13b*): The Immobilines were devised primarily to overcome the three difficulties described previously. In so doing, casting gradients in narrow cylinders became cumbersome. Thus, the use of horizontal slabs with plastic backing sheets for first dimension focusing became popular. One-stage gradient SDS gels (without a stacking gel) were used originally in the second dimension, but the preparation of 'seamless', two-stage discontinuous SDS slab gels has popularized the Laemmli system for use in a horizontal second dimension.

The strategy for using this two-dimensional system is to run a horizontal first dimension using the instructions given in *Protocol 11*, then equilibrate the gel strip and its plastic backing sheet in SDS-containing buffers. The second dimension SDS pore gradient gel is cast on to a plastic backing sheet with or without a stacking gel. The equilibrated first dimension strip is laid, at right angles, on top of the second dimension gel, at the cathodal (−) end. Electrophoresis proceeds in the second dimension until the dye front reaches the bottom of the gradient resolving gel.

Special techniques

Peptide mapping (see *Protocol 14*)

The technique of peptide mapping using gel electrophoresis is frequently used for two different purposes: comparison of protein relationships and amino acid sequence analysis. Peptide mapping to deduce the possible relationships between two proteins involves enzymatic cleavage of the proteins and comparison of the resultant fragment patterns. Using proteolytic enzymes which cleave at specific residues, identical proteins will have the same patterns when the cleavage products are subsequently separated by gel electrophoresis; related proteins will have at least one fragment in common. As originally practiced, peptide mapping required enzymatic digestion in solution, which in turn, required purification of the proteins in question. The Cleveland technique [23], however, is performed on a single component of a mixture of proteins resolved by SDS–

polyacrylamide gel electrophoresis. Accordingly, the strategy for performing Cleveland mapping experiments is to separate a mixture of proteins by SDS gel electrophoresis. Following electrophoresis, the gel is lightly stained, and the band containing the protein of interest is excised. This band is re-equilibrated in SDS-containing buffers and inserted into the well of a stacking gel on top of a second SDS separating gel. The excised block is overlaid with a solution containing proteolytic enzymes that are active in the presence of SDS (e.g. V8 protease, chymotrypsin or papain). The target protein and the enzyme are driven into the stacking gel by the current. The power is then turned off, and digestion proceeds in the gel. Following digestion, the current is reapplied, and the cleavage products are resolved in the separating gel.

A second, newer application of Cleveland mapping relates to protein sequence analysis using sequencing methods from protein blots. One of the frequent difficulties encountered in protein sequencing by Edman degradation is that the N-terminal α-amino group is blocked and becomes unreactive to Edman's reagent. In this case, sequence is obtained from an internal peptide fragment and a corresponding synthetic oligonucleotide is prepared. The oligonucleotide can then be used to probe a cDNA library to identify the nucleic acid sequence encoding the complete protein.

Immunoelectrophoresis (see *Protocol 15*)

Immunoelectrophoresis entails separating the components of the test sample by electrophoresis in agarose and identification of the separated components by immunodiffusion against antibodies. The binding of specific antibodies to their reactive antigens usually results in the formation of an immune precipitate. The evidence of a precipitate allows immunoelectrophoresis to be used for two purposes: (i) to identify an unknown protein band using a specific antiserum, or (ii) to determine whether an unknown antiserum contains antibodies to a particular antigen contained in a mixture of proteins.

An insoluble immune precipitate forms when the molar concentrations of antigen and antibody are approximately equal (known as the equivalence point); precipitation does not occur in situations of antigen or antibody excess (see the *Antibody Applications* volume of this series). When the concentration of antibodies and antigens is high, the precipitate will be visible in the electrophoresis gel to the naked eye. Lower concentrations will require staining.

The strategy for performing immunoelectrophoresis experiments is to separate the proteins by one-dimensional electrophoresis in agarose, then to cut a trough in the gel parallel to the lane, adjacent to the separated bands. The trough is filled with the antibodies which diffuse through the agarose. Immune precipitates form in an arc-shaped pattern at the points where diffused antigens and antibodies meet and specific antibody–antigen interactions take place. The center-point of the arc corresponds to the location of the antigen band.

Owing to the requirement for diffusion following the electrophoresis step, immunoelectrophoresis is necessarily performed in large-pored gels, preferably agarose. Consequently, some band sharpness is sacrificed. The gel techniques almost always use nondenaturing buffer systems because antibodies are most frequently raised against native proteins. The commonest of these techniques uses a continuous barbital buffer system but other nondenaturing systems are also known.

Rocket immunoelectrophoresis (see *Protocol 16a*)
Rocket immunoelectrophoresis (also called Laurell rocket) is a rapid, convenient means to quantify, without purification, the levels of a particular antigen contained in a mixture of proteins. The technique is based on using the current to drive the antigen into an antibody-containing gel. The antigen–antibody complexes precipitate in the gel, at and where, the equivalence point is reached. Precipitation begins at the lateral borders of the sample wells and continues in the direction of antigen electrophoretic migration to a peak, forming a 'rocket'-shaped pattern. The height of the rocket is directly proportional to the antigen concentration. By merely applying a set of known concentrations of antigen in wells on the same gel, a calibration curve is readily obtained.

In order for the method to work, steps must be taken to ensure that the antigens and antibodies meet. The antibodies must remain stationary by conjugation to the gel matrix [24] or be driven rapidly toward the antigen. The latter approach is most common. For this, the electrophoresis is conducted at pH 8.6. At this pH, most proteins will move toward the anode (see *Protocol 9*), but this is the isoelectric point of IgG. If an agarose having a high electroendosmotic characteristic is used, then the electroendosmosis will drive the antibody toward the antigen.

Crossed immunoelectrophoresis (see *Protocol 16b*)
Crossed (also called two-dimensional) immunoelectrophoresis is a combination of immunoelectrophoresis (*Protocol 15*) and rocket immunoelectrophoresis (*Protocol 16a*). The strategy for crossed immunoelectrophoresis is to separate proteins using a continuous buffer system in the first dimension. The lane is cut out and run at right angles into a second-dimensional gel containing multi-specific antibodies.

The technique can be used to quantify the amounts of a single antigen present in a complex biological fluid [25] or can be used to analyze the antigenic composition of biological materials [26].

Choice of methods

The choice between using charge- and size-based approaches is necessarily influenced by many factors. The most important are prior information about the components of the mixture and purpose of the separation. Obviously, the most delicate separations are those where protein function must be preserved. These are primarily preparative, protein interaction and zymography experiments. In these cases, electrophoresis under conditions which seek to maintain the protein in its 'native' form is preferred. Frequently these requirements result in a compromise between conditions optimal for the separation and conditions optimal for protein function. When maintenance of the 'native' conditions is not possible, renaturation procedures with or without gel elution steps can be added.

Separation of proteins under denaturing conditions is used for several reasons. The most common purposes are: molecular weight estimation, studies of the subunit structure and conformation of proteins, and peptide mapping applications. In other cases, separations under denaturing conditions are used for preparations whose solubilization requires detergents, polar solvents, chaotropic ions or sulfhydryl agents. The denaturing agents most frequently used in electrophoresis experiments are SDS and urea.

Protocol 9. **Tris-glycine polyacrylamide gels**

Reagents

Acrylamide/bisacrylamide stock – 30 g acrylamide ⚠, 0.8 g bisacrylamide ⚠ made up to 100 ml with distilled water
5% Ammonium persulfate (w/v) in distilled water
Running (tank) buffer – 6.0 g Tris base, 28.8 g glycine to 1 l (adjust to pH 8.3 if necessary)
5× Sample buffer solution – mix 15.5 ml of 1 M Tris-HCl (pH 6.8) with 2.5 ml of 1% (w/v) Bromophenol Blue, 7 ml of distilled water and 25 ml of glycerol
5× Separating gel buffer – 22.7 g Tris base, 140 μl TEMED ⚠ in 100 ml of 0.48 M HCl (adjust pH to 8.9 if necessary)
5× Stacking gel buffer – 7.48 g Tris base, 280 μl TEMED ⚠ in 100 ml of 0.48 M HCl (adjust to pH 6.7 if necessary)

Equipment

Electrophoresis cell, preferably with cooling capacity (e.g. BioRad Protean, Pharmacia Multiphor, miniProtean or equivalent)
Hamilton syringe or automatic pipettor (e.g. Gilson)
Power supply (constant voltage)
Well-forming comb

Procedure

1 To perform Tris-glycine electrophoresis in a 10%T, 2.66%C polyacrylamide gel, choose one of the formats given in *Protocol 5a* or *c* and assemble the plates and backing sheets or tubes.

2 To make 12 ml of gel solution, mix:
- 4.0 ml of acrylamide/bisacrylamide solution

Notes

This procedure will take approximately 5 hours.

① 12 ml is the appropriate volume for many standard electrophoresis cells (such as the BioRad Protean II with separate casting stand) using a 0.75 mm spacer.

② Samples should be added to sample buffer such that a final concentration of 1× sample buffer is achieved.

- 2.4 ml of 5× separating gel buffer
- 5.48 ml distilled water. ①

3 De-gas according to *Protocol 5a*, step 5 and note 7. Add 120 µl of persulfate solution, mix by gentle swirling and cast the gel by following the instructions in *Protocol 5a*. If cylindrical gels are used, follow *Protocol 5c*. [1]

4 Following polymerization, prepare 10 ml of a 4%T, 2.66%C stacking gel by mixing:

- 1.33 ml acrylamide/bisacrylamide solution
- 2 ml of 5× stacking gel buffer
- 6.57 ml distilled water.

Add 100 µl of ammonium persulfate. Remove the water layer, fill the cassette to the top and insert the well-forming comb.

5 When the stacking gel is polymerized, remove the comb and transfer the gel cassette from the casting stand to the upper reservoir, securing it with the off-center cams.

6 Fill the upper and lower electrolyte reservoirs with running buffer.

7 Load the samples into the sample wells using a Hamilton syringe or pipettor. In the case of the latter, the sample can be loaded gently by turning the volume adjuster slowly clockwise. Then switch on the current. The cathode, negative electrode, is at the top; separation proceeds toward the positive electrode. ②③④⑤

Maintenance of pH at 6.7–6.8 is important for the proteins to stack properly. If the sample is not buffered to pH 6.8 by addition of sample buffer alone, neutralize the sample or dialyze it first against 1× sample buffer.

③ Glycerol-containing sample buffer will increase the density of the sample, causing it to sink to the bottom of the well. Filling the well more than half-full with sample is not recommended.

④ The cooling capacity of the system limits the applied voltage. For one gel slab, electrophoresis at 100 V (constant voltage) should take about 3 h with tap water as the coolant.

⑤ When running two slabs, they must both have the same height and composition. Otherwise, as a parallel electrical circuit, the current will be shunted proportionally to the path of least resistance; the gels will run at different rates.

⑥ If fragile gels are not supported by backing sheets, pry the plates apart, remove the top plate and cover the gel with a nylon mesh (a nylon stocking will do). Invert the gel and ease the gel from the remaining plate with a spatula, allowing the gel to fall on to the mesh. The mesh will be used to carry the gel through subsequent fixing and staining steps.

8 Electrophoresis is complete when the Bromophenol Blue tracking dye reaches within 1 cm of the bottom of the gel. Pry the plates apart, cut the slab at the seam between the stacking and separating gel and discard the stacking gel. Remove the separating gel for subsequent fixing, staining, blotting, etc. Fragile gels (pores larger than about 6%T) tear easily and should be supported by a nylon 'hammock'.⑥

Pause point

[1] See note 8 to *Protocol 5a*.

Protocol 10a. Isoelectric focusing in rehydratable polyacrylamide slab gels

Reagents

Ampholyte solution – when a narrow range (e.g. pH 5–7) is needed, prepare a water solution which is 4% (v/v) pH 5–7 Ampholine, 10% (v/v) glycerol. If a wide range (e.g. pH 3–10) is needed, substitute a blend of 0.8% (v/v) 3–10 Ampholine, 3.2% 5–7 Ampholine, 10% glycerol ①②

Anolyte (+) solution 1 M H_3PO_4 ⚠

Catholyte (−) solution 1 M NaOH ⚠

pI markers (Appendix B)

Equipment

Horizontal electrophoresis cell with cooling system (e.g. Pharmacia Multiphor, E-C 1001 or equivalent)

Paper electrode wicks or gel cubes

Power supply, preferably constant wattage

Sample applicator

Surface pH electrode (Pharmacia or Microelectronics) or pH meter

Procedure

1 Use a 5%T, 2.66%C rehydratable gel prepared according to the instructions given in *Protocol 6.* ③

2 Rehydrate the gel by soaking it in ampholyte solution according to *Protocol 6*. To be economical of ampholyte solution and to ensure that the gel does not overhydrate, use of a rehydration cassette is recommended. (Directions given in *Protocol 6.*) ④

3 When rehydration is complete, cover the gel with Gelbond or plastic wrap until use to prevent drying. Before use, remove excess ampholyte

Notes

This procedure will take approximately 5 hours.

① For buffer electrofocusing, a premixed 47 component buffer mixture is commercially available (Polysciences).

② This protocol is written for use with ampholytes of the trademark 'Ampholine' from Pharmacia. Characteristics of carrier ampholyte solutions vary as the range of the gradient varies, from manufacturer to manufacturer and from lot to lot from a single manufacturer. *Always refer to the manufacturer's instructions for use.*

③ This protocol is written for a horizontal 10 cm × 10 cm ×

from the gel surface by drawing a piece of polyester backing sheet across the surface in the manner of a windscreen-wiper.

4 Place the rehydrated gel on the cooling platen with a water layer between the plastic backing sheet and the platen to ensure good contact.⑤

5 Prepare electrode wicks by soaking at least eight thicknesses of filter paper (Whatman No. 1 or equivalent) in H_3PO_4 anolyte and NaOH catholyte. For separations which require long focusing times, the gradients can be stabilized by changing the catholyte to histidine or lysine depending on the range of ampholytes used [32]. Place the wicks on the gel surface beneath the electrodes. An alternative to paper wicks is the use of gel cubes (described in Chapter III).

6 Prefocus the gel for 30 min at 100 V, constant voltage.

7 Apply the sample anywhere on the surface. If applicable, loading a protein near its isoelectric point saves focusing time.⑥

8 Focus the gel for at least 3 h at 1.5 W, constant power. If the gel is kept at 15°C, set the power supply to limit at 1000 V. Determining when to end the run must be assessed experimentally; load samples or markers in separate lanes at different locations between the electrodes. Focusing is complete when they are all aligned.⑦⑧⑨

9 Measure/estimate the pH gradient if desired. Either: (i) use a surface pH electrode, or (ii) cut off a 10 mm wide strip of gel, slice it sequentially in 2 mm increments, incubate each slice in 2 ml distilled water to allow the ampholines to leach out, then read the pH, or (iii) use pI marker proteins.

0.5 mm slab gel with a 9 cm interelectrode distance. Dried gels on plastic backing sheets can be cut to size prior to rehydration.

④ Refs 12, 27 and 28 give additional information.

⑤ Isoelectric points are temperature-dependent. Uniform temperature across the complete gel is essential.

⑥ Sample application is detailed elsewhere [29]. Samples in small volumes can be applied directly to the surface (about 5 μl), on a paper wick or using an applicator 'mask' (10–50 μl), a 'containment collar' made from a disposable pipette tip (>50 μl) [30] or in large volumes (>100 μl), using a trough [31] obtainable from Pharmacia.

⑦ The electrical resistance of the gel changes during focusing. A power supply which can be set at constant power will save time by continuously upregulating the voltage in response. Set it to limit by voltage at a level for which the heat exchanger can maintain the temperature. If in doubt, it is best to choose a lower voltage and longer focusing time.

⑧ When markers loaded at opposite sides of the gel are aligned, all but very large proteins will have focused. Band sharpening is sometimes improved by increasing the voltage at the end. If cooling capacity is sufficient, increase to 2000 V for the last 10 min.

⑨ For some applications, it may be necessary to increase the suggested catholyte concentrations five-fold over those given in the protocol [32,13].

Protocol 10b. Isoelectric focusing in agarose gels

Reagents

Ampholytes (see note 2 of *Protocol 10a*)
Anolyte (+) solution 1 M H_3PO_4 ⚠
Catholyte (−) solution 1 M NaOH ⚠
Isogel agarose (FMC), Agarose-IEF (Pharmacia) or equivalent specifically designated for electrofocusing
pI markers (Appendix B)

Equipment

Boiling water bath or microwave oven
Electrophoresis plate
Horizontal electrophoresis cell with cooling system (e.g. Pharmacia Multiphor, E-C 1001 or equivalent)
Leveling table
Paper electrode wicks or gel cubes
Power supply, preferably constant wattage
Sample applicator
Waterproof tape

Procedure

1. To prepare a 10 ml (10 cm × 10 cm × 1 mm) agarose-isoelectric focusing gel, assemble the plate and plastic backing sheet according to *Protocol 5d.*①
2. Weigh out 0.05 g Isogel agarose, add 10 ml of water and heat in a boiling water bath or microwave oven.②③ (See also notes 3 and 4 to *Protocol 4.*)
3. Allow to cool to 60–70°C. Add 150 μl Ampholine 3.5–10, 50 μl Ampholine 2.5–4 and 50 μl Ampholine 5–8 and gently mix by swirling.

Notes

This procedure will take approximately 3 hours.

① Agarose gels thinner than about 1 mm tend to dry out rapidly during the run.

② Agaroses with different electroendosmosis properties are available. To minimize electroendosmosis-induced gradient drift, use only those agaroses specifically designated for isoelectric focusing (e.g. Agarose-IEF, Pharmacia or Isogel, FMC).

③ A 0.5% (w/v) gel will accommodate very large proteins (about 500 kDa). For smaller proteins, increase the agarose concentration to 1%.

4 Pour on to plate. If necessary, spread the liquid to the edges of the plate using a pipette tip. When the solution has gelled, put the gel into the refrigerator for 1 h. [1]

5 Place the gel on the cooling platen, with a water layer between the plastic backing sheet and the platen to ensure good contact.

6 Prepare electrode wicks (or gel cubes as described in Chapter III) by soaking at least eight thicknesses of filter paper (Whatman No. 1 or equivalent) in anolyte and catholyte. Place them on the gel surface beneath the electrodes.

7 Load the sample. For application and placement, see notes 6 and 8 to *Protocol 10a*. The prefocusing step can be omitted in this protocol.

8 Focus the sample at 3 W, constant power (see note 8 to *Protocol 10a*) until complete.

9 Measure the gradient if desired (see *Protocol 10a*, step 9).

Pause point

[1] Agarose gels containing ampholytes can be stored overnight in a humidified chamber at 4°C.

Protocol 10c. **Isoelectric focusing in ultrathin gels (pH 5–7)**

Reagents

Ampholyte solution – 4% (v/v) 5–7 Ampholine, 10% (v/v) glycerol in distilled water (see note 2 of *Protocol 10a*) ①

Anolyte (+) solution 1 M H_3PO_4 ⚠

Catholyte (−) solution 1 M NaOH ⚠

pI markers (Appendix B)

Equipment

Horizontal electrophoresis cell with cooling system (e.g. Pharmacia Multiphor, E-C 1001 or equivalent)

Paper electrode wicks (Whatman No.1 or equivalent)

Power supply, constant wattage

Sample applicator

Procedure

1 Prepare a 10 cm × 10 cm rehydratable ultrathin 5%T, 2.66%C rehydratable gel according to the flap technique described in *Protocol 5b*. Use a polythene sandwich bag for a 60 μm spacer or a strip of Parafilm as a 120 μm spacer as required.

2 Rehydrate the gel by soaking it in ampholyte solution.

3 When rehydration is complete, cover the gel with Gelbond or plastic wrap until use to prevent drying. Before use, be certain to remove any excess ampholyte solution from the gel surface. This is best done by drawing a piece of polyester backing sheet across the surface in the manner of a windscreen-wiper.

4 Place the rehydrated gel on the cooling platen with a water layer

Notes

This procedure will take approximately 1 hour.

① For suggestions regarding wide pH range gradients, see ref. 6.

② Temperature control is critical at these high voltages. See notes 5 and 7 to *Protocol 10a*.

③ Very high voltage isoelectric focusing (up to 1700 V/cm of interelectrode distance, as opposed to 190 V/cm here) has been reported [33].

④ See notes to *Protocol 10a* regarding use of markers and determining when focusing is complete. Estimating the gradient of ultrathin gels is best done using pI marker proteins.

between the plastic backing sheet and the platen to ensure good contact. Set the platen temperature to 5°C.②

5 Prepare electrode wicks by soaking at least eight thicknesses of filter paper in anolyte and catholyte. Place them on the gel surface beneath the electrodes.

6 Prefocus the gel for 5 min at 1 W, constant power. Set the power supply to limit at 250 V.

7 Apply the sample on a coarse filter paper wick (see note 6 to *Protocol 10a*).

8 Focus the gel for 5 min at 2 W, constant power. Set the power supply limit at 500 V.

9 Focus the gel for 10 min at 4 W, 900 V limit.

10 Remove the wicks and focus for an additional 5 min at 8 W, 1600 V limit.

11 Focus for 2 min at 12 W, 1900 V limit.③④

12 Measure the pH gradient, if desired.④

Protocol 11. Isoelectric focusing using Immobilines

Reagents

Acrylamide/bisacrylamide stock – 29.1 g acrylamide ⚠, 0.9 g bisacrylamide ⚠ made up to 100 ml with distilled water
Immobilines (Pharmacia)
Persulfate solution – 5% (w/v) ammonium persulfate in distilled water
TEMED ⚠
pI markers (Appendix B)
Rehydration solution ①②

Equipment

Electrophoresis plates
Gradient maker (e.g Isco, Pharmacia or equivalent)
Horizontal electrophoresis cell with cooling system (e.g. Pharmacia Multiphor, E-C 1001 or equivalent)
Magnetic stirrer and stirring bar
Paper electrode wicks or gel cubes
Peristaltic pump (e.g. Labcanco, Pharmacia or equivalent)
Power supply, preferably constant wattage
Sample applicator
Silanized plastic backing sheets
Spacers

Procedure

1 To perform isoelectric focusing in a linear pH 4–10, 4–7 or 7–10 immobilized gradient, prepare a rehydratable polyacrylamide slab gel of 3%T, 3%C as follows. (Recipes courtesy of Pharmacia Biotech, with permission.) Assemble the plates and spacers to form the gel mold cassette.

2 Prepare the heavy (acidic) solution and light (basic) solutions by mixing according to *Table 2*. Neutralize. ③

Notes

This procedure will take approximately 11 hours.

① Rehydration solutions vary depending on the nature of the separation. Charge-carrying electrolytes are not necessary because the Immobilines make the gel conductive. Their presence [e.g. 0.2–1% (w/v) carrier ampholytes] aids in maintaining the sample proteins in solution and reduces the separation time [17]. A good general rehydration

3 Vacuum de-gas (*Protocol 5a*, step 5 and note 7) the heavy and light solutions. Gently, mix in 5 μl TEMED and 50 μl persulfate solution to each then cast the linear gradient using the instructions in *Protocol 7*.

4 Polymerize the gel at 50°C for 1 h.

5 Wash the gel three times in 200 ml of distilled water for 20 min. [1]

6 Slice the gel into strips 5–10 mm wide (5 mm maximum if this is the first dimension of a two-dimensional separation – see *Protocol 13b*). A batch of 40–80 strips can be prepared simultaneously from this recipe.

7 If the gel strips have been stored, prepare a rehydration cassette as in note 3 of *Protocol 6*. Use a 0.5 mm thick spacer which is slightly larger than the dimensions of the strip (wider for rehydrating multiple strips simultaneously). Insert the gel strip into the cassette and fill it with rehydration solution. Rehydrate the gel at room temperature for 1 h (viscous solutions such as those containing glycerol or urea require 2 h). ④ [2]

8 Place the rehydrated gel on the cooling platen with a water layer between the plastic backing sheet and the platen, to ensure good contact.

9 Load the samples. See notes 6 and 8 to *Protocol 10a*.

10 Set the cooling platen to 15°C and run the system at 300 V for 1 h. Remove the applicator wicks, masks or troughs then focus for 6–7 h at 5 W, constant power. See notes 7 and 8 of *Protocol 10a* for duration of focusing. ⑤⑥

solution is 0.5% (w/v) carrier ampholytes. Match the ampholyte range to the range of the Immobiline gradient.

② If possible, inclusion of the sample in the rehydration solution is recommended. This can decrease focusing time and avoid concentration steps during sample preparation.

③ Polymerization will proceed unevenly across the gel – faster at the basic end. To ensure consistent polymerization, neutralize the heavy solution with 1 M NaOH, and the light solution with 1.5 M acetic acid. Check the pH of the solutions by spotting 2–5 μl on to pH paper. The pH gradient must be restored by washing out the NaOH and acetic acid after the gel has polymerized.

④ The rehydration cassette limits the thickness to a uniform 0.5 mm. The acidic end rehydrates faster than the basic end.

⑤ Anolytes and catholytes are not necessary.

⑥ Power setting is determined by the number of strips focused simultaneously; 5 W is appropriate for strips with a combined width of 5–10 cm. If the cooling system is sufficiently efficient, use a power setting which limits at a maximum voltage of 5000 V (3000 V if the rehydration solution contained ampholytes).

⑦ pH gradients cannot be measured by pH electrodes in Immobiline gels. Use pI markers instead (Appendix B).

⑧ Coomassie Blue-stained gels cannot be fully destained due to binding by the Immobilines. A faint background will remain.

11 Measure the pH gradient if desired. (7)

12 Fix and stain the gel. If soluble stain is to be used, use *Protocol 17a*. If carrier ampholytes were included in the rehydration solution, use *Protocol 17c*. (8)

Pause points

[1] The use of precast rehydratable Immobiline slab gels (Pharmacia) is advantageous. Alternatively, prepare several rehydratable slabs simultaneously, or one large rehydratable gel, which can be cut into smaller slabs as needed. This recipe will make a slab 10 cm wide × 40 cm long × 0.5 mm thick. After washing with distilled water (step 5), soak the gels in 2% (v/v) glycerol for 30 min and allow to air dry overnight. The dried gels can be stored in plastic bags at −20°C for 1 year [34]. Longer-term storage is not recommended.

[2] For rehydration solutions not containing labile samples, rehydration overnight is possible.

Table 2. Formulations for immobilized linear pH gradients

		pH 4–10		pH 4–7		pH 7–10	
		Heavy (μl)	Light (μl)	Heavy (μl)	Light (μl)	Heavy (μl)	Light (μl)
Immobiline	pK 3.6	0.551	—	0.289	0.151	0.271	0.45
	pK 4.6	—	0.57	0.55	0.369	—	—
	pK 6.2	0.227	0.25	0.225	0.75	—	—
	pK 7.0	0.45	0.244	—	0.135	0.189	0.162
	pK 8.5	0.167	0.78	—	—	0.175	0.175
	pK 9.3	—	0.179	—	0.438	—	0.140
Acrylamide/bisacrylamide solution		1.0ml	1.0 ml	1.0 ml	1.0 ml	1.0 ml	1.0 ml
Glycerol		2.0 ml	0.3 ml	2.0 ml	0.3 ml	2.0 ml	0.3 ml
Water		7.5 ml	7.5 ml	7.5 ml	7.5 ml	7.5 ml	7.5 ml

For formulations of linear gradients having other pH ranges and for formulations of nonlinear Immobiline gradients see ref. 16.

Protocol 12. SDS–polyacrylamide gels

Reagents

Acrylamide/bisacrylamide stock – 30 g recrystallized acrylamide ⚠, 0.8 g bisacrylamide ⚠ made up to 100 ml with distilled water
5% (w/v) Ammonium persulfate in distilled water
Molecular weight marker proteins (Appendix B) ①
Running (tank) buffer – 3.027 g Tris-base, 14.41 g glycine, 1.0 g SDS. Adjust pH to pH 8.3 if necessary and make up to 1 l with distilled water
SDS sample buffer – 0.98 g Tris-HCl, 2.0 g SDS, 7.5 ml glycerol, 5 ml 2-mercaptoethanol. Adjust pH to 6.8 and make up to 100 ml with distilled water ②
5× Separating gel buffer – 226.9 g Tris base, 1.25 ml TEMED ⚠, 5.0 g SDS. Adjust pH to 8.8 and make up to 1 l with distilled water ③
5× Stacking gel buffer – 9.85 g Tris-HCl, 0.125 ml TEMED ⚠, 0.5 g SDS. Adjust to pH 6.8 and make up to 100 ml with distilled water
TEMED ⚠

Equipment

Power supply, constant voltage
Vertical slab gel apparatus

Procedure

1 To cast a vertical 10%T, 2.66%C, 12 ml SDS slab gel, first assemble the plates and spacers to form the gel mold cassette (*Protocol 5a*).
2 Make 12 ml of gel solution by mixing (gently to avoid frothing):
- 4.0 ml acrylamide/bisacrylamide solution
- 2.4 ml 5× separating gel buffer
- 5.5 ml distilled water.

3 Vacuum de-gas the mixture (*Protocol 5a*, step 5 and note 7). Add 100 μl ammonium persulfate.

Notes

This procedure will take approximately 5 hours.

① Molecular weight markers in 1× SDS sample buffer should be included in every run. For best results, load them in lanes at both the left and right sides of the slab. For the majority of reduced proteins to be separated in 10%T, single density gels, use a mixture of five or six proteins ranging from 10 to 100 kDa. For nonreduced proteins in a 5%T single density gel, use markers of 70–500 kDa.

4 Pour or pipette the gel mixture between the plates. Fill the cassette high enough to allow for 1 cm of stacking gel beneath the teeth of the well-forming comb.

5 Carefully pipette a 3 mm layer of water or water-saturated butanol on top of the acrylamide solution. Allow to polymerize. [1]

6 When the separating gel is polymerized, pour off the water layer and prepare 10 ml of the 3%T, 2.66%C stacking gel by mixing:
- 1.0 ml of acrylamide/bisacrylamide solution
- 2.0 ml 5× stacking gel buffer
- 6.9 ml distilled water
- 50 μl ammonium persulfate solution
- 50 μl TEMED.

7 Fill the space between the plates to the top, then insert the well-forming comb.

8 When the stacking gel is polymerized, remove the comb and transfer the gel cassette from the casting stand to the upper reservoir, securing it with the off-center cams.

9 Fill the upper and lower electrolyte reservoirs with tank buffer and check that there are no leaks.

10 Ensure that the sample is completely complexed with SDS, by incubating sample+sample buffer for 2 min at 100°C. Allow to cool to room temperature and add 1–2 μl of 0.1% (w/v) Bromophenol Blue per 100 μl. Load the samples into the sample wells then apply the current. The negative electrode is at the top; separation proceeds toward the positive electrode. ④

② SDS gel electrophoresis can be performed with or without reduction of disulfide bridges. For nonreducing conditions, omit 2-mercaptoethanol from the sample buffer. 2 mM dithiothreitol can be substituted for 2-mercaptoethanol since it has no unpleasant odor.

③ Use only SDS specifically designated for electrophoresis and use a single manufacturer, as different sources of SDS can give different results!

④ See notes 4–6 of *Protocol 9* for additional advice on loading, running and ending the electrophoresis.

Pause point

[1] See note 8 to *Protocol 5a*.

Protocol 13a. Two-dimensional electrophoresis with carrier ampholytes using a cylindrical first-dimensional gel and a vertical second-dimensional slab gel

Reagents

5% (w/v) Ammonium persulfate in distilled water
Ampholines 5–7 and 3–10 (Pharmacia). See note 2 to *Protocol 10a*
First dimension acrylamide/ bisacrylamide stock – 30 g recrystallized acrylamide ⚠, 1.8 g bisacrylamide ⚠ made up to 100 ml with distilled water
First dimension anolyte (+) solution 10 mM H_3PO_4 ⚠
First dimension catholyte (−) solution 20 mM NaOH ⚠①
First dimension overlay solution – 2 ml 9 M urea, 40 μl ampholine 5–7, 5 μl ampholine 3–10①
10% (v/v) NP-40 solution in distilled water
SDS equilibration buffer – 0.98 g Tris-HCl, 2.0 g SDS, 7.5 ml glycerol, 5 ml 2-mercaptoethanol. Adjust to pH 6.8 and make up to 100 ml with distilled water. Add 5 mg Bromophenol Blue②③
0.3% (w/v) SeaKem gold agarose (FMC Bioproducts) in SDS equilibration buffer
Second dimension acrylamide/ bisacrylamide stock – 30 g recrystallized acrylamide ⚠, 0.8 g bisacrylamide ⚠ made up to 100 ml with distilled water
Second dimension running (tank) buffer. 3.027 g Tris-base, 14.41 g glycine, 1.0 g SDS. Adjust to pH 8.3 and make up to 1 l with distilled water②
5× Second dimension separating gel buffer – 226.9 g Tris base, 1.25 ml TEMED ⚠, 5.0 g SDS. Adjust to pH 8.8 and make up to 1 l with distilled water②
5× Second dimension stacking gel buffer – 9.85 g Tris-HCl, 0.125 ml TEMED ⚠, 0.5 g SDS. Adjust to pH 6.8 and make up to 100 ml with distilled water②

Equipment

Electrophoresis plates
Gel tubes (2 mm × 140 mm)
Gradient maker (e.g. Isco or Pharmacia)
Magnetic stirrer and stirring bar
Peristaltic pump (e.g. Labcanco or Pharmacia)
Power supply, preferably constant wattage
Silanized plastic backing sheets
Spacers
Tube gel apparatus
Vertical slab gel apparatus

Procedure

1 Cast 12 first-dimensional electrofocusing gels in cylinders 140 mm long and 2 mm inner diameter as follows.
(i) Mark the outside of the cylinder 10 mm from the top. Seal the bottom with Parafilm (see note 1 of *Protocol 5c*).
(ii) Prepare the gel solution by mixing:
- 5.5 g urea
- 1.33 ml first dimension acrylamide/bisacrylamide solution
- 2 ml NP-40 solution
- 1.87 ml distilled water
- 0.4 ml pH 5–7 Ampholines
- 0.1 ml pH 3–10 Ampholines
- 15 μl TEMED.

Dissolve urea completely and allow to warm to room temperature. Vacuum de-gas the solution (*Protocol 5a*, step 5 and note 7), then add 20 μl of ammonium persulfate solution.
(iii) Mix gently to avoid frothing.
(iv) Place the tubes upright in a rack and fill them to the mark with gel solution. See step 2 of *Protocol 5c*. ④
(v) Overlay the gel solution with a 50 μl distilled water layer and allow to polymerize for at least 1 h. Replace the distilled water layer with whatever sample buffer is to be used. ⑤ [1]

2 Fill the bottom of the electrophoresis cell with anolyte. Remove the Parafilm from the bottom of the tubes and insert them into the electrophoresis cell. Displace any air bubbles trapped at the bottom of the cylinders by jetting anolyte with a 'J' tube (a bent Pasteur pipette).

Notes

This procedure will take approximately 36 hours.

① For a pH 3.5–10 first-dimensional gel, gradient drift will cause the basic end to migrate off the gel. It will end at about pH 8.5. Gradient drift can be retarded by changing the catholyte to 20 mM histidine [32]. For gradients with narrower ranges, see ref. 13.

② Second-dimensional solutions are identical to those for the Laemmli technique (*Protocol 12*).

③ Equilibration solution is Laemmli sample buffer with Bromophenol Blue added.

④ At this concentration, urea is near its solubility limit. If urea crystallizes in the gel, keep the gel at a slightly warmer than ambient temperature or reduce the amount of urea to 5 g and increase the amount of distilled water to 2 ml.

⑤ Conductivity discontinuities can be caused if the distilled water layer is left for long periods. The gel, when covered with sample buffer or sealed with agarose, can be stored overnight.

⑥ The sample and the top of the gel are protected from the extreme pH of the catholyte by using an overlay solution.

⑦ For nonequilibrium first-dimensional gels make the following modifications to the procedure:
(i) Do not prefocus the gel.
(ii) The bottom of the cylindrical gel will be in the NaOH solution, which will be the anolyte, and the top of the cylinder with the sample will be in the H_3PO_4 solution, which will be the catholyte.
(iii) Run the system at 1000 V (constant voltage) for 4 h.

3 Replace the sample buffer layer with 50 μl of first-dimensional overlay solution.⑥

4 Fill the space from the top of the overlay solution to the top of the tube with catholyte solution then fill the upper electrolyte reservoir with catholyte.⑦

5 Prefocus the gel at 200 V (constant voltage) for 1 h.

6 Turn off the current, remove the catholyte and overlay solution from the tubes. Load the samples in a volume of less than 50 μl. Cover the samples with overlay solution, replace the catholyte. Run the system at 500 V (constant voltage) for 16 h, then at 1000 V for 2 h.⑧

7 Remove the gels from the tubes. ⑨ [2]

8 Measure the pH gradient using a microelectrode or see step 9 of *Protocol 10a*.

9 Mark the gel such that the positive end can be distinguished from the negative end following equilibration (and storage). ⑩

10 Equilibrate the first-dimensional gel by incubating for 30 min (with periodic gentle mixing) at room temperature in a test tube containing 10 ml of SDS equilibration buffer. ⑪ [2]

11 Cast a 5–20% linear gradient, 1.5 mm thick SDS separating gel according to the instructions given in *Protocols 7* or *8* and *12*. If the entire two-dimensional procedure is performed without interruption, this should be ready when the first-dimensional electrophoresis is finished.

⑧ A time-voltage product of 10 000 V h is used here. If your cooling system cannot maintain the temperature below 15°C, reduce the applied voltage and increase the time to achieve 10 000 V h.

⑨ The simplest way to remove the gel is by slowly tightening a vice against the tube and breaking the glass. Pick up the gel using a gloved hand and place on Parafilm. Alternatively extrude the gel with water pressure. Connect a plastic hose to a water-filled syringe on one side and the gel tube on the other. Secure the hose to the syringe with a hose clamp or wire. Decrease the pressure on the plunger as the gel begins to move. This step requires artistry (lots of it!), practice on blank gels!

⑩ Slice off a few millimeters of the positive end with a diagonal cut *or* inject ink into one end *or* keep the gels in a container on which the orientation is marked. Gels with SDS-containing samples will have an identifying SDS-bulge at the positive end.

⑪ Omitting the equilibration step, suggested by some workers to minimize protein loss, is not good practice. Equilibration time is a trade-off between complete complexing of proteins with SDS + mercaptoethanol and protein diffusion. If diffusion or loss is a problem, reduce the equilibration time to 15–20 min.

⑫ Stacking gels in the second dimension are not always used. In that case, the second dimension gradient is frequently 10–20%T instead of 5–20%T [35] .

⑬ Use of molecular weight markers for two-dimensional gels is discussed elsewhere [36]. Cast a 5%T, 2.66%C

12 Cast a 1 cm stacking gel. If the entire two-dimensional procedure is performed without interruption, this should be started when the equilibration is started.⑫

13 After equilibration, place the first-dimensional gel on to a piece of Parafilm, straighten it out with a spatula, and aspirate any residual equilibration buffer. Slide the gel cylinder on to the top of the slab cassette so that it rests in the space between the plates. Using a thin plastic spatula, starting from one end of the first-dimensional gel, gently push the cylindrical gel between the plates until it contacts the top of the stacking gel at all points. Even though the cylindrical gel is 0.5 mm wider in diameter than the space between the plates, it will slide in easily due to presence of SDS from the equilibration.

14 If marker slices are to be used, push them now on to the top of the stacking gel, at both ends of the first-dimensional gel.⑬

15 Prepare a solution of 0.3% (w/v) SeaKem Gold agarose in equilibration buffer. Melt the agarose and allow it to cool to about 55°C. See notes 3 and 4 of *Protocol 4* for instructions on melting agarose. Pipette enough agarose into the gel slab cassette to seal the first-dimensional gel in place.⑭

16 Fill the cassette to the top with running buffer, insert the cassette into the electrophoresis cell and electrophorese the slab at 100–200 V (constant voltage) until the dye front is within 1 cm of the bottom of the slab.

17 See notes 4–6 of *Protocol 9* for running and terminating electrophoresis in the second dimension.

cylindrical gel (2 mm i.d.) containing a set of markers. Substitute dithiothreitol for mercaptoethanol as the reducing agent. Extrude the gel, equilibrate it as you would a first-dimensional gel, cut it into 3 mm slices and freeze the slices individually. When loading the isoelectric focusing gel on to the second-dimensional slab, thaw out two slices and place one at either end of the first dimension.

⑭ Several different agaroses are available for sealing the first-dimensional gel in place such as LGT (BioRad) Isogel and Sea Plaque (FMC). Stacking gels made from Isogel instead of polyacrylamide have also been reported [37]. Some agaroses leave streaking artifacts following silver staining, which can be minimized by inclusion of iodoacetamide in the equilibration solution [38]. SeaKem Gold agarose (FMC) is recommended for gels to be visualized with silver stain.

⑮ Some procedures require solubilization of the proteins in SDS, which will mask their pIs for the first dimension. However, inclusion of NP-40 in the first-dimensional gel displaces SDS previously bound to the proteins.

Pause points

1 See note 5.

2 The first-dimensional gels can be stored indefinitely at −70°C before or after equilibration. Storage after equilibration, when the urea is allowed to diffuse out, is recommended. Some authors find loss of protein if the gels are stored in the first dimension tubes.[39]

Protocol 13b. Two-dimensional electrophoresis using horizontal Immobiline slabs in the first dimension and horizontal SDS slabs in the second dimension

Reagents

Dithiothreitol solution – 200 mg/ml DTT in distilled water (prepare fresh)

Equilibrating solution – 36 g recrystallized urea, 30 ml glycerol, 2 g SDS, 10 ml 0.5 M Tris-HCl, pH 6.8, 5 mg Bromophenol Blue, made up to 100 ml in distilled water ① ②

Solutions for first-dimensional Immobiline gels according to *Protocol 11*, except rehydrating solution

Solutions for second-dimensional SDS gels according to *Protocol 12*, except sample buffer solution

Rehydrating solution – 0.5% (w/v) carrier ampholytes of the same range as the Immobiline gradient, 0.5% (v/v) NP-40, 0.2% (w/v) dithiothreitol made up in 8 M recrystallized urea ③

Equipment

Electrophoresis plates
Gradient maker (e.g. Isco or Pharmacia)
Horizontal electrophoresis cell with cooling system (e.g. Pharmacia Multiphor, E-C 1001 or equivalent)
Magnetic stirrer and stirring bar
Paper electrode wicks or gel cubes
Peristaltic pump (e.g. Labcanco or Pharmacia)
Power supply, preferably constant wattage
Sample applicator
Silanized plastic backing sheets
Spacers

Procedure

1 Prepare and run a rehydratable Immobiline gel according to the instructions given in *Protocol 11*, but use the rehydrating solution given here. [1]

Notes

This procedure will take approximately 16 hours.

① Urea and glycerol are present in the equilibration buffer to facilitate complete transfer of proteins from first to

2 Assemble the plates (21 × 13 cm), plastic backing sheets and U-form spacer (21 × 13 cm overall dimension, 0.7 mm thick and 1 cm wide) to cast a linear 10–15%T, 2.66%C 'seamless' pore gradient, Laemmli second dimension. Follow the general instructions given in *Protocols 7* and *12*. This gel is cast upside down, with the stacking gel at the bottom of the cassette. See note 4 for the characteristics of this slab gel. ④⑤

3 Prepare 6.2 ml of the heavy solution (10%T) by mixing:
- 2.08 ml of acrylamide/bisacrylamide solution
- 1.24 ml of 5× separating gel buffer
- 2.25 ml of distilled water
- 0.62 ml of glycerol.

4 Prepare 6.2 ml of the light solution (15%T) by mixing:
- 3.18 ml of acrylamide/bisacrylamide solution
- 1.24 ml of 5× separating gel buffer
- 1.77 ml of distilled water.

5 Prepare 5.1 ml of stacking gel solution by mixing:
- 1.0 ml of 5× stacking gel buffer
- 0.5 ml acrylamide/bisacrylamide solution
- 3.0 ml distilled water
- 2.0 ml glycerol
- 25 μl TEMED.

6 Vacuum de-gas (*Protocol 5a*, step 5 and note 7) the solutions prepared in steps 3–5 and chill them, along with the cassette to 4°C.

second dimension. Without them, first-dimensional Immobiline gels become glued to the second-dimensional slab, which in turn, will become distorted.

② For gels to be silver stained, include 260 mM iodoacetamide in the equilibration buffer. This will reduce 'streaking' caused by unreacted DTT remaining in the second-dimensional gel.

③ Carrier ampholytes are not required in the rehydration solution. First-dimensional separation usually proceeds faster with them.

④ This protocol uses a first dimension of 10 cm in length and a second dimension of 19.5 cm long × 11 cm wide × 0.7 mm thick. Of the 19.5 cm length, the separating gel is 16 cm long (volume=12.4 ml) and the stacking gel is 3.5 cm (volume=2.7 ml). A space of 0.5 cm is left at the top of the cassette for a water layer.

⑤ Other popular sizes for horizontal two-dimensional separations use first-dimensional gels of 15 and 18 cm. A 3.5 cm stacking gel is still used but the separating gels are adjusted to 12 cm and 15 cm, respectively. The corresponding gradient volumes for the second dimensions are 9.3 ml and 11.6 ml, respectively.

⑥ Work quickly from this point to ensure that the second-dimensional slab gel is poured before polymerization begins.

⑦ A gel strip containing molecular weight markers is prepared in the same way as for the cylindrical gels in *Protocol 13a*.

7 Add 44 μl of persulfate solution to the heavy and light solutions, 25 μl to the stacking gel solution.⑥

8 Pipette 2.7 ml of stacking gel into the cassette, pour the heavy solution into the mixing chamber of the gradient maker and the light solution into the other chamber. Pump the gradient on top of the stacking gel and cover the gradient with a 3 mm distilled water layer.

9 Mix 0.5 ml of DTT solution with 10 ml of equilibrating solution and incubate the first-dimensional strip for 15 min at room temperature. This is best done in a stoppered test tube placed horizontally on a rocking platform. Repeat for another 15 min cycle. If iodoacetamide is required, add it to the second cycle of equilibration (see note 14 of *Protocol 13a*). [2]

10 Place the second-dimensional SDS gel on the cooling platen with the stacking gel toward the cathode (−).

11 Saturate the electrode wicks (at least eight thicknesses of Whatman 3MM filter paper) in SDS running buffer. Place them on to the gel surface such that they cover about 15 mm of gel. Place the equilibrated first-dimensional gel strip face down on the stacking gel, immediately adjacent to the cathode wick. Make sure that excess equilibrating solution is removed from the first-dimensional gel strip. When molecular weight markers are used, place them alongside the first-dimensional strip. See note 1 of *Protocol 12* and note 13 of *Protocol 13a*.⑦

12 Set the cooling system for 10–15°C and run at 200 V (constant voltage)

until the Bromophenol Blue dye front reaches the interface of the stacking and separating gels.

13 Turn off the power and remove the cathode wick and first-dimensional strip. Replace the cathode wick so that it covers the area previously occupied by the first-dimensional gel strip.

14 Run the system at 500 V (constant voltage) until the dye front reaches the end of the separating gel.

Pause points

[1] Rehydratable Immobiline gels can be stored prior to electrophoresis for up to 1 year at −70°C [34].

[2] After electrophoresis, the first-dimensional gels can be stored indefinitely at −70°C. This is best done following the first cycle of equilibration. The second cycle is performed immediately prior to electrophoresis in the second dimension.

Protocol 14. **Peptide mapping by the Cleveland technique**

Reagents

Destaining solution – 50 ml absolute methanol ⚠, 100 ml glacial acetic acid ⚠, 850 ml distilled water

Enzyme solution – equilibration buffer containing 10% (v/v) glycerol, 0.5 μg/ml purified V-8 protease and 0.001% (w/v) Bromophenol Blue ①②

Equilibration buffer – 0.125 M Tris-HCl, 0.1% (w/v) SDS, 1 mM EDTA. Adjust to pH 6.8

Staining solution – 1.1 g Coomassie Brilliant Blue R-250, 500 ml methanol, 100 ml glacial acetic acid, 500 ml distilled water

Equipment

Electrophoresis apparatus
Light box
Power supply
Rocking platform
Staining dish

Procedure

1 Cast a native or denaturing slab electrophoresis gel appropriate for the sample that contains the protein of interest and perform the electrophoresis. Use a gel of 1.5 mm thickness. ③④⑤

2 Following electrophoresis, incubate the gel, at room temperature for 30 min in staining solution and for 30 min in destaining solution. Change the destaining bath at least once during this period.

3 Lay the destained gel on a light box and excise the band of interest. Trim the slice to a width that will easily fit into the sample well of the second gel.

Notes

This procedure will take approximately 8 hours.

① The Cleveland technique is limited to the few proteolytic enzymes (e.g. papain, chymotrypsin and *Staphylococcus aureus* V-8 protease) which can be obtained in high purity and also retain activity in SDS-containing buffers.

② Enzyme solution can be stored frozen in small aliquots. Concentration will vary according to specific activity.

③ For fragments to be visualized by conventional staining, start with about 10 μg of protein per band of interest, less for other detection methods.

④ Isoelectric focusing gels are rarely used for the first gel due

4 Incubate the gel slice in equilibration buffer for 30 min at room temperature. [1]

5 Cast the second gel. The separating gel should be a 15%T, 2.66%C uniform concentration SDS gel or a 15–20% linear gradient SDS gel, of 1.5 mm thickness. Gels of 15 or 20%T are made by adjusting the amounts of acrylamide/bisacrylamide solution and water in the recipes for *Protocols 7* or *12*. The separating gel should be cast to be 1 cm shorter than that described in *Protocol 12* and the difference made up with stacking gel.⑥

6 Carefully push the gel slices to the bottom of the sample wells in the stacking gels with a spatula and add 20 μl of equilibration buffer containing 20% glycerol to the wells. This should cover the slices. If not, add a little more. Carefully layer 10 μl of enzyme solution on top using a Hamilton syringe. Without disturbing the layer of enzyme solution, carefully fill the sample wells to the top with running buffer.⑦

7 Run the system at 100 V, constant voltage, until the Bromophenol Blue tracking dye reaches the interface between the stacking gel and separating gel.

8 Turn off the current and allow the digestion to take place for 20–30 min at room temperature.

9 Run the system at 200 V, constant current, until the dye front reaches within 1 cm of the bottom of the separating gel.

to staining difficulties (see *Protocol 17*).

⑤ When a choice of %T is available for the first gel, choose the one with the largest pores.

⑥ The second gel should be ready to run at the end of the incubation in equilibration buffer, or immediately on thawing of the frozen gel slices.

⑦ Save at least one lane for molecular weight markers and one lane for a blank gel slice. This is needed to identify the position of the enzyme and any potential contaminants in the pattern.

Pause point

[1] Gel slices can be stored frozen before or after equilibration.

Protocol 15. Immunoelectrophoresis

Reagents

Agarose ②
Barbital buffer (0.05 M) – 1.84 g of barbital ⚠ and 10.3 g of sodium barbital ⚠ made up to 1 l with distilled water. Adjust to pH 8.6 ①

Equipment

Boiling water bath or microwave oven
Electrophoresis plates
Filter paper
Horizontal electrophoresis cell
Leveling table
Light box
Power supply

Procedure

1. Cast a 10 ml (10 cm × 10 cm × 1 mm) 1% (w/v) agarose gel in barbital buffer according to *Protocol 5d*.
2. Allow the gel to set and use a scalpel to cut the template shown in *Figure 8* from the gel. Wells can be cut in place of an applicator wick.
3. Remove the gel from the plugs by gentle lifting with the point of a needle or with gentle suction. To remove the troughs, invert the agarose and pull them away from the main body of the gel with fine-tipped forceps. ③
4. Place the gel on the cooling platen with the troughs parallel to the field.

Notes

This procedure will take approximately 36 hours.

① Barbital buffer is stored at 4°C for as long as the pH is maintained. In jurisdictions where barbital is a controlled substance, use Tris-tricine (39.2 g Tris-HCl, 17.2 g tricine, 4.2 g calcium lactate). Adjust to pH 8.6 and make up to 1 l with distilled water.

② The agarose recommended for immunoelectrophoresis is of medium electroendosmosis (EEO approx. −0.13).

③ The main body of the gel must not separate from the support during removal of the plugs and troughs. Otherwise, antibodies will seep under the gel rather than diffuse through it during the diffusion step.

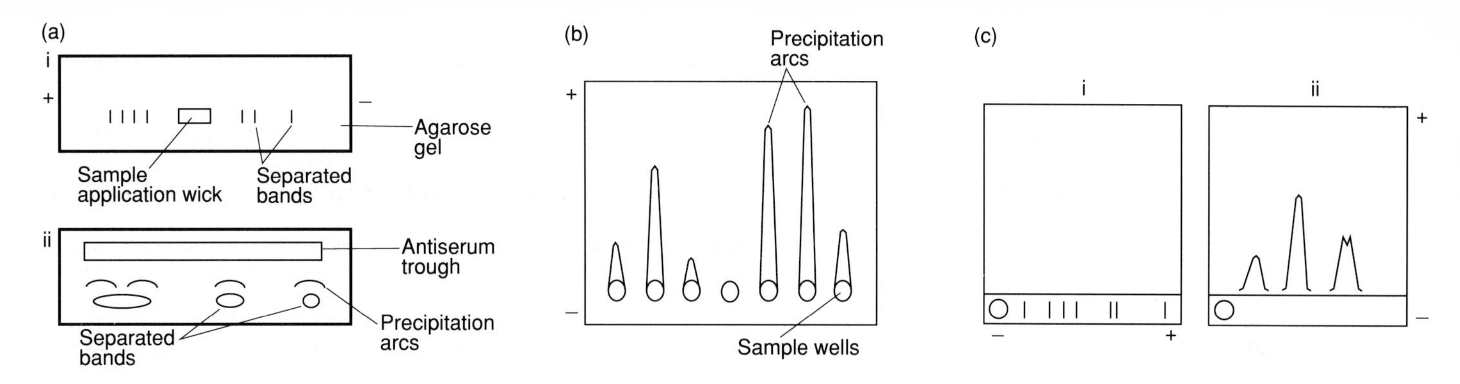

Figure 8. (a) Immunoelectrophoresis: i, electrophoretic separation; ii, stained immunoelectrophoresis pattern showing antibody reactivity to four of six proteins. (b) Rocket immunoelectrophoresis: stained pattern showing antibody recognition for six of seven proteins. (c) Crossed immunoelectrophoresis: i, immunoelectrophoresis gel showing first electrophoretic separation gel fused to second antibody-containing gel; ii, stained pattern showing antibody reactivity to four of seven proteins.

5 Fill the buffer reservoirs with 0.05 M barbital buffer and connect the reservoirs to the gel with electrode wicks (at least eight thicknesses of Whatman 3MM filter papers pre-wet with the barbital buffer).

6 Fill the sample wells with sample made up in barbital buffer containing Bromophenol Blue tracking dye (50 μg/ml).④

7 Run the system at 100 V (constant voltage) until the tracking dye reaches the end of the troughs at the anodal (+) side.

8 At the end of the run, turn off the power, fill the troughs with antiserum and place the gel in a humidor (see pause point 1 to *Protocol 5d*) at 4°C and leave for 20–24 h.⑤

9 Remove and examine the gel on a light box for signs of visible precipitation arcs. If none appear, continue the incubation for another 24 h. If the immunoprecipitate is visible to the naked eye, staining is unnecessary. If staining is required, it will be necessary to first remove the unreacted antiserum from the agarose. To do this follow steps 10–12.

10 Remove the remaining antiserum from the troughs.

11 Fill the wells with water and cover the plate with Whatman 3MM filter paper and a pad of absorbent tissue. Place a weight (such as a glass plate) on top and blot the liquid out of the gel. Wash the gel in 0.1 M NaCl for 10 min. Repeat the pressing/washing cycle once or twice more.⑥

12 Fix and stain the gel with Coomassie Blue according to *Protocol 17a*.

④ For horizontal electrophoresis on an open gel, the sample wells must be filled to the top in order to ensure even current flow. Immunoelectrophoresis can also be run as a 'submarine' system, in which case the sample must be made denser with glycerol [add glycerol to the sample to a concentration of about 15% (v/v)].

⑤ The time required for the immunodiffusion step depends on the concentration of antibodies and antigens and is necessarily determined experimentally.

⑥ All proteins in the gel will stain. Pressing and washing cycles are necessary to remove proteins that did not immunoprecipitate. Otherwise, a high background is obtained.

Protocol 16a. **Rocket immunoelectrophoresis**

Reagents

Agarose ①
Barbital buffer (0.05 M) – 1.84 g of barbital ⚠ and 10.3 g of sodium barbital ⚠ made up to 1 l with distilled water. Adjust to pH 8.6. See note 1 to *Protocol 15*

Equipment

Boiling water bath or microwave oven
Electrophoresis plates
Filter paper
Horizontal electrophoresis cell
Leveling table
Light box
Power supply

Procedure

1 Cast a 10 ml (10 cm × 10 cm × 1 mm) 1% (w/v) agarose gel in barbital buffer according to *Protocol 15* with the following modifications: (i) Before pouring, allow the agarose solution to cool to 56°C in a water bath. Add the appropriate amount of antiserum and mix by gentle inversion. Re-equilibrate the agarose solution to 56°C, then pour the gel; (ii) Do not cut the troughs.②

2 Allow the gel to set and cut sample wells as shown in *Figure 8b* from the gel.

3 Remove the gel from the plugs gently (see *Protocol 15*).

Notes

This procedure will take approximately 5 hours.

① Agarose having a high EEO characteristic (about −0.25) is used for these procedures.

② The amount of antibody used in the gel is necessarily determined experimentally. With a strong hyperimmune system, a good rule of thumb is to start with 200–300 μl of serum per 10 ml gel.

③ These procedures will not work if the target antigens' isoelectric points are the same or higher than IgG. In that case, the antigens can be covalently modified to change their pI [40].

4. Place the gel on the cooling platen with sample wells at the negative side.
5. Fill the buffer reservoirs with 0.05 M barbital buffer and connect the reservoirs to the gel with electrode wicks (at least eight thicknesses of Whatman 3MM filter papers pre-wet with the barbital buffer).
6. Fill the sample wells with test sample made up in barbital buffer. Make a serial dilution (in barbital buffer) of antigen of known concentrations and fill the wells with a volume equal to that of the test samples.③
7. Run the system at 100–120 V (constant voltage) for 1 h, stop the run and examine the plate on a light box. If the peaks of the rockets are small, with little difference between the standards, continue the run for another hour. Again remove and examine the gel for signs of visible precipitate. If the immunoprecipitate is visible to the naked eye, staining is unnecessary. If staining is required, complete steps 11 and 12 of *Protocol 15*.

Protocol 16b. **Crossed immunoelectrophoresis**

Reagents

Agarose ①
Barbital buffer (0.05 M) – 1.84 g of barbital ⚠ and 10.3 g of sodium barbital ⚠ made up to 1 l with distilled water. Adjust to pH 8.6. See note 1 to *Protocol 15*

Equipment

Boiling water bath or microwave oven
Electrophoresis plates
Filter paper
Horizontal electrophoresis cell
Leveling table
Light box
Power supply

Procedure

1 Prepare the first-dimensional plate as if performing *Protocol 15*, but do not cut out troughs.

2 Load the samples and run the first dimension under the same conditions as given in *Protocol 15*.

3 While the first dimension is running, prepare the second dimension, antibody-containing gel solution according to *Protocol 16a*, but do not pour the gel. Keep the gel solution at 56°C until ready for use. ②③

4 At the end of the run, cut the lanes out (with the Gelbond backing sheets attached if Gelbond was used), using a scalpel and straight edge. ④

Notes

This procedure will take approximately 19 hours.

① Agarose having a high EEO characteristic (about −0.25) is used for these procedures.

② The amount of antibody used in the gel is necessarily determined experimentally. With a strong hyperimmune system, a good rule of thumb is to start with 200–300 μl of serum per 10 ml gel.

③ These procedures will not work if the target antigens' isoelectric points are the same or higher than IgG. In that case, the antigens can be covalently modified to change their pI [40].

5. Prepare the plate which will be used for the second dimension according to *Protocols 4* and *5d*, and transfer the first-dimensional strip to the second-dimensional plate, aligning it along one edge. See *Figure 8c*.
6. Pour the second-dimensional gel solution on to the plate, ensuring a good contact with the first-dimensional strip.
7. When the molten gel has completely set, place it on the cooling platen with the first-dimensional strip at right angles to the field.
8. Fill the electrode compartments with barbital buffer (step 5 of *Protocol 16a*).
9. Attach the first-dimensional strip to the cathodal buffer wick.
10. Connect the anode buffer wick to the opposite side of the gel and run the system at 20–30 V (constant voltage) for 16 h.
11. Examine the gel on a light box for signs of visible precipitate. If the immunoprecipitate is visible to the naked eye, staining is unnecessary. If staining is required, complete steps 11 and 12 of *Protocol 15*.

④ To minimize diffusion, cutting the gel strip from the first dimension and casting the second-dimensional gel should be performed immediately after the end of the first dimension run.

V DETECTION OF PROTEINS IN GELS

Introduction

Proteins, apart from the few colored ones, are invisible to the naked eye in the amounts used for electrophoresis. Consequently, indirect methods are required to deduce their presence and mark their location in the gel. The most common means of protein detection is staining, in which a soluble chromophore binds directly to the protein in the gel matrix [1]. The best of the soluble stains has a detection sensitivity of around 1 μg in a tight band or spot. A variation of staining which has added sensitivity uses the stain as a colloidal dispersion [2,3]. A second variation is the use of metal (most frequently silver) stains, in which metal grains are precipitated on the proteins at or near the gel surface. Silver stains have been reported to detect proteins in the nanogram range. In other applications, the proteins can be labeled pre- or post-electrophoresis with detectable groups (sometimes called 'reporters'). The most common reporters are radioactive, and their detection sensitivity can be in the subnanogram range. This remains the most reliable approach for detecting very small amounts of protein. One alternative approach combines staining and radioactive reporters to create a radioactive stain [25].

When the amount of protein in the gel is high, the bands or spots will tend to merge with each other. Hence, protein resolution improves as the protein load in the gel decreases. Resolution is therefore bounded, on the one hand by the sensitivity of the detection method and on the other hand by the range of concentrations of the various proteins in the sample. As a result, fractionation of the sample prior to electrophoresis is frequently used to increase resolution. This strategy has particular appeal for blood serum samples where approximately 50% of the protein mass is albumin [4].

The ability to detect a spot or band in the gel is determined by the concentration of bound stain or metal grain per unit volume of gel. Resolution and detection sensitivity can therefore be enhanced by making the bands 'tighter' – most frequently done by decreasing the pore size of the gel and decreasing the run time (e.g. ultrathin gels, *Protocol 10c*). Resolution and detection sensitivity can be preserved by limiting diffusion of the proteins during staining. This is usually done by acid- or alcohol-fixation of the proteins immediately following electrophoresis and maintenance of these conditions during the staining. Some procedures combine fixation and staining in a single step [5]. As an alternative to chemical fixation, diffusion can be prevented by drying the gel rapidly using microwaves [6, 7].

Methods available

Soluble and colloidal stains (see *Protocol 17*)

The most common soluble stains are Coomassie Blue, Amido Black and Ponceau Red. Gels can be fixed, stained and then destained; this will serve the majority of applications (*Protocol 17a*). However, while straightforward and relatively rapid, it produces background staining of the gel. Attempts to remove the background by diffusion destaining will release stain from the protein. Soluble stains are also used to identify the carbohydrate and lipid moieties of glycoproteins and lipoproteins. To stain for the carbohydrate moieties of glycosylated proteins, the Periodate–Schiff reaction (Thornton's modification [8] of the Fairbanks procedure [9]) is most commonly used. A more sensitive fluorescent technique is available which uses periodate oxidation and borohydride reduction to attach a dansyl group [10].

Staining for lipid material is inefficient and is complicated by the fact that the stains are necessarily hydrophobic. Hence the published procedures are not entirely satisfactory. Two approaches are possible; the first is a procedure which prestains the protein with Nile Red before electrophoresis [11], while the second is adapted from an

References

1. Wilson, C.M. (1983) *Meth. Enzymol.* **91**, 366–247.
2. Neuhoff, V., Stamm, R. and Eibl, H. (1985) *Electrophoresis*, **6**, 427–448.
3. Patestos, N.P., Fauth, M. and Radola, B.J. *Electrophoresis*, **9**, 488–496.
4. Gersten, D.M., Khirabadi, B.S., Kurian, P., Ledley, R.S., Mahany, T., Ramey, E.R. and Ramwell, P.W. (1980) *Biochem. J.* **191**, 869–872.
5. Reisner, A.H., Nemes, P. and Bucholtz, C. (1975) *Analyt. Biochem.* **64**, 509–516.
6. Gersten, D.M., Zapolski, E.J. and Ledley, R.S. (1983) *Analyt. Biochem.* **129**, 57–59.
7. Gersten, D.M., Zapolski, E.J. and Ledley, R.S. (1985) *Electrophoresis*, **6**, 191–192.
8. Thornton, D.J., Carlstedt, I. and Sheehan, J.K. (1994) *Meth. Mol. Biol.* **32**, 119.
9. Fairbanks, G., Steck, T.L. and Wallach, D.F.M. (1971) *Biochemistry*, **10**, 2606–2617.
10. Gander, J.E. (1984) *Meth. Enzymol.* **104**, 447–451.
11. Greenspan, P. and Gutman, R.L. (1993) *Electrophoresis* **14**, 65–68.
12. Noble, R.P. (1968) *J. Lipid Res.* **9**, 693–700.

older procedure which uses Oil Red or Sudan Black to stain a dried gel [12].

Coomassie Blue G-250 [2] or Acid Violet 17 (CI 42,650) [3] can be used as a colloidal dispersion. The particles of the colloid are too large to enter the interior of the gel. Rather, the stain partitions on to the protein exposed at the gel surface. This results in a background gel which is perfectly clear. Although the Coomassie Blue process (*Protocol 18*) takes longer than the procedures using soluble stains, detection sensitivity has been reported to be increased 10-fold; detection of nanogram levels of protein has been reported [13].

Metal stains (see *Protocol 19*)

For the detection of nanogram levels of protein without radioactivity, metal stains are used. The most common metal staining techniques use silver; silver stains have two major chemistries based on silver nitrate and silver diamine. Due to the proven sensitivity of silver staining, considerable effort has been directed toward simplifying and standardizing the technique. However, the reaction mechanism(s) is/are still not clear (see refs 14 and 15 for a detailed discussion of silver staining). Where silver staining is required, the procedure modified from Guevara *et al.* [16] is recommended. Many different procedures are available [15], and rapid stains are available in 'kit' form (e.g. Accurate Chemical and Scientific Corp.). The procedure given in *Protocol 19* will serve all applications and is particularly suit-

13. Neuhoff, V., Arold, N., Taube, D. and Erhardt, W. (1988) *Electrophoresis*, **9**, 255–262.
14. Lognonne, J.-L. (1994) *Cell Mol. Biol.* **40**, 41–45.
15. Rabilloud, T., Vuillard, L., Gilly, C. and Lawrence, J.-L. (1994) *Cell Mol. Biol.* **40**, 57–75.
16. Guevara, J., Johnston, D.A., Ramagli, L.S., Martin, B.A. Capitello, S. and Rodriguez, L.V. (1982) *Electrophoresis*, **3**, 197–205.
17. Gorg, A., Postel, W. and Gunther, S. (1988) *Electrophoresis*, **9**, 531–546.
18. Chamberlain, J.P. (1979) *Analyt. Biochem.* **98**, 132–135.
19. Ludlow, J.W., Yuen, C.K.L. and Consigli, R.A. (1985) *Analyt. Biochem.* **145**, 212–215.
20. Luthi, U. and Waser, P.G. (1965) *Nature*, **205**, 1190–1191.
21. Laskey, R.A. (1980) *Meth. Enzymol.* **65**, 363–371.
22. Smith, A.G., Phillips, C.A., Hahn, E.J. and Leacock, R.J. (1985) *AAS Photo Bull.* **39**, 8–15.
23. Smith, A.G., Phillips, C.A. and Hahn, E.J. (1985) *J. Imaging Technol.* **11**, 27-32.
24. Phillips, C.A., Smith, A.G. and Hahn, E.J. (1986) in *Synthesis and Applications of Isotopically Labelled Compounds '85* (R.R. Mucino, ed.), pp. 189–193. Elsevier, Amsterdam.
25. Zapolski, E.J., Gersten, D.M. and Ledley, R.S. (1982) *Analyt. Biochem.* **123**, 325–328.
26. Radola, B.J. (1983) in *Electrophoretic Techniques* (C.F. Simpson and M. Whittaker, eds), pp. 101–118. Academic Press, London.

able for thick (1.5 mm), gradient (10–20%T) gels as are used in two-dimensional applications. Two-dimensional gels stained with silver sometimes have streaking artifacts which are absent from conventionally stained gels, or radioactive gels. For a discussion of these see ref. 17.

Radioactive proteins

Proteins can be rendered radioactive metabolically (using ^{3}H, ^{14}C or ^{35}S-amino acids) or by covalent modification (most commonly with ^{3}H, ^{14}C, ^{125}I, ^{131}I, ^{203}Hg). Autoradiography and fluorography are the least expensive means of detecting radioactive proteins, particularly those labeled with weak β-emitters. To perform autoradiography, an X-ray film is placed in direct contact with the gel slab. For autoradiography of high energy emitters, the gel need only be placed next to a sheet of X-ray film and stored at −70°C in a light-tight container. Wrap the gel in plastic wrap to avoid wetting the film. Use a film cassette or a cardboard folder to sandwich the gel and film tightly together. For gels frozen immediately following electrophoresis, acid/alcohol fixation is unnecessary; the pattern will not diffuse in a frozen gel. If subsequent staining is not required and the gel is to be retained, the protein pattern can be 'fixed' in the gel by rapid drying using microwaves [6].

Weak emitters (particularly tritium) are detected by radiofluorography (often called fluorography; see *Protocol 20*) because their emis-

27. Rodriguez, L.V., Gersten, D.M., Ramagli, L.S. and Johnston, D.A. (1993) *Electrophoresis*, **14**, 628–637.
28. Waterborg, J.H. and Matthews, H.R. (1994) *Meth. Mol. Biol.* **32**, 163–167.
29. Boffhard, H.F. and Datyner, A. (1977) *Analyt. Biochem.* **82**, 327–333.

Protocols provided

17a. *Staining gels following acid/alcohol fixation*
17b. *Rapid staining using Coomassie Brilliant Blue*
17c. *Staining isoelectric focusing gels containing carrier ampholytes*
17d. *Staining for glycoproteins*
17e. *Staining for lipoproteins*
18. *Colloidal Coomassie Blue staining*
19. *Silver staining of gels*
20. *Radiofluorography*

sion is susceptible to self-absorption by the gel itself; the β-particles never reach the film emulsion. Tritium is only detectable by fluorography. ^{14}C and ^{35}S are on the borderline for detection by autoradiography, but they are almost always detected by fluorography. Fluorography makes use of the fact that some film emulsions are far more sensitive to the visible and near visible wavelengths than to the emissions themselves. For the weaker emitters (e.g. ^{3}H, ^{14}C, ^{35}S) the gel is impregnated or sprayed with a scintillant, which will emit light in response to bombardment with β-particles. Impregnation of the gels is reported to be more sensitive than spraying. Both water-soluble sodium salicylate, which is more convenient [18], and water-insoluble PPO, which is more sensitive [19], scintillants have been used.

Fluorography is mandatory for detecting low energy β-particles. It is often also used to increase sensitivity and reduce exposure time when detecting the stronger emitters (strong β-particles such as ^{32}P and electromagnetic emitters). Simply place the gel and film on a tungstate intensifying screen which will serve as a scintillant. When intensifying screens are used (Dupont Lightning Plus, Kodak X-omatic, or their equivalents), dry the gel first (*Protocol 27*) then place the film between the gel and the screen. Many screens come already mounted in a light-tight holder (cassette). To reduce exposure time further, three strategies can be used. One is to expose the film and scintillant at −70 to −80°C [20]. A second method is to preflash the film. This uses light to initiate the nucleation centers upon which the silver precipitate will form [21]. A third method is to 'hypersensitize' the X-ray film by preincubation at 65°C in a gas bag containing 8% H_2 and 92% N_2 [22–24]. This method can be combined with exposure at low temperature to achieve even greater time-saving [23].

An alternative to autoradiography and fluorography for radioisotopes (e.g. ^{125}I) is fractionation of the gel followed by electronic detection. As self absorption is not a problem for gels containing such isotopes, these can be sliced sequentially and the slices counted in a gamma or liquid scintillation counter.

Choice of methods

The choice of detection method depends primarily on the amount of protein used in the gel (see Protein loading guide, *Table 1,* p. 6). The vast majority of applications are served by staining with soluble stains. For detecting submicrogram amounts, colloidal stains are the least expensive approach but require several days. Metal staining can be labor intensive and can lead to serious artifacts. The most serious is that proteins vary widely in their 'stainability'. Nonetheless, metal staining has proven sensitivity which rivals that of radioactive detection.

Fluorography and autoradiography are the least expensive means of detecting radioactive proteins because electronic counting instruments are not required. The trade-off for economy is that the procedures are inefficient, requiring long exposure times. Phosphoimagers are now available for counting weak β-particles in a two-dimensional format, but due to their high cost, their use is not widespread (e.g. BioRad and Molecular Dynamics). Electronic detection is the method of choice for the higher energy emitters, offering speed and better counting efficiency. Electronic detectors for higher energy emitters are widely available.

Protocol 17a. **Staining gels following acid/alcohol fixation**

Reagents

Destaining solution – same as staining solution without the Coomassie Brilliant Blue R-250
Fixing solution – 300 ml absolute methanol ⚠, 34.5 g sulfosalicylic acid, 115 g trichloroacetic acid ⚠ made up to 1 l in distilled water
Staining solution – 250 ml absolute ethanol ⚠, 80 ml glacial acetic acid, 1.0 g Coomassie Brilliant Blue R-250 (also called Serva Blue R-250) made up to 1 l with distilled water

Equipment

Filter paper
Light box
Rocking platform or wrist shaker
Staining dish

Procedure

1 Place the gel in a shallow dish containing 10 times the gel volume of fixing solution at room temperature. For unbacked gels about 1 mm thick, incubate for at least 1 h with rocking.① [1]

2 Filter the staining solution immediately before use as Coomassie Blue rarely dissolves fully.

3 Decant the fixing solution and replace with 10 volumes of staining solution. Incubate with occasional rocking for at least 1 h.①

4 Decant the staining solution and replace with destaining solution. Incubate with constant rocking. At least five changes of destaining

Notes

This procedure will take approximately 6 hours.

① Double the incubation times for gels with backing sheets and for those 2–3 mm thick.

② Staining is complete when the color intensity looking through the gel (on a light box) is equal to that surrounding the gel. Destaining is complete when the gel is as clear as fresh destaining solution.

③ Destaining can be accelerated by absorbing the stain as it is released from the gel. With the second change of destaining solution, place a small skein of bleached wool (about 6 g), or a small sack of activated charcoal, into the destaining dish.

solution will be required to destain the gel fully. If more than a few hours elapses between the time the gel is destained and the time it will be dried, photographed, etc., it should be stored in destaining solution containing 0.005% (w/v) Coomassie Brilliant Blue R-250.② ③ [2]

Pause points

[1] The gel can be left in fixing solution without harm.
[2] The gel can be left in destaining solution between changes.

Protocol 17b. **Rapid staining using Coomassie Brilliant Blue**

Reagents

Staining solution – 0.4 g Coomassie Brilliant Blue G-250 in 1 l of 3% (w/v) perchloric acid

Equipment

Filter paper
Light box
Rocking platform or wrist shaker
Staining dish

Procedure

1 Filter the staining solution. The staining solution will be orange but the stain will turn blue when it binds to the protein. Stain the gel for 45 min at 37°C in a shallow dish containing a volume of staining solution equal to 160% of the volume of the gel.

2 Decant the stain. Destaining is not required. The gel can be stored in 3.5% perchloric acid containing 0.005% (w/v) Coomassie Blue G-250.

Notes

This procedure will take approximately 1 hour.

Protocol 17c. **Staining isoelectric focusing gels containing carrier ampholytes** ① ②

Reagents

Destaining solution – same as staining solution without the Coomassie Blue G-250

Fixing solution – 20% (w/v) trichloroacetic acid (TCA) ⚠ in distilled water

Staining solution – 0.5 g Coomassie Blue G 250 dissolved in 250 ml methanol ⚠, 650 ml distilled water, 100 ml glacial acetic acid ⚠

Equipment

Filter paper
Light box
Rocking platform or wrist shaker
Staining dish

Procedure

1 Fix the gel in TCA fixing solution for 1 h at room temperature,during which time the ampholytes will leach out of the gel. (See pause point 1 of *Protocol 17a.*)

2 Remove the TCA from the gel by incubating it in destaining solution for 1 h at room temperature.

3 Stain and destain with the above staining and destaining solutions using the same procedure given in *Protocol 17a*, steps 3 and 4.

Notes

This procedure will take approximately 6 hours.

① Staining ultrathin isoelectric focusing gels requires close attention to timing. One very important advantage of ultrathin gels is that detection sensitivity is greater than that in conventional gels due to the 'tightness' of the bands (see Chapter III). For gels 60–120 μm thick, fix 5–10 min, rinse in destaining solution 30 sec, stain 5–10 min, destain 5–10 min [26].

② The use of fixatives containing methanol at room temperature is not advised for isoelectric focusing gels containing carrier ampholytes. The ampholytes, which will stain with Coomassie Blue R-250, will be precipitated in the gel and are not easily removed.

Protocol 17d. **Staining for glycoproteins**

Reagents

Fixing solution – 50% (v/v) ethanol ⚠ in distilled water
Metabisulfite solution – 0.5% (w/v) sodium metabisulfite in 10 mM HCl. *Make immediately before use*
Periodate solution – 1 ml periodic acid, 3 ml glacial acetic acid ⚠, 97 ml distilled water. *Make immediately before use*
Preserving solution – 7.5 ml absolute methanol ⚠, 5 ml glacial acetic acid made up to 100 ml with distilled water
Schiff's reagent – mix 1 g basic fuchsin (Sigma) with 200 ml boiling distilled water. Cool to 50°C, filter, then add 20 ml 1 M HCl. Allow to cool then add 1 g potassium metabisulfite. Store in the dark overnight, add 2 g activated charcoal, shake and decant. Store in the dark

Equipment

Filter paper
Light box
Rocking platform or reciprocating shaker
Staining dish

Procedure

1 Fix the gel in fixing solution for 30–60 min.① [1]

2 Wash in water 10 min on a rocking platform or reciprocating shaker. ②

3 Incubate for 30 min in periodate solution.

4 Wash through six changes of water for at least 5 min each. ③

Notes

This procedure will take approximately 7 hours.

① Incubation times given are for unbacked gels about 1 mm thick. Double the times for backed gels with only one open surface. For 2–3 mm gels, double the times.

② Shake the gel on a rocking platform or reciprocating shaker, not a wrist shaker. [8]

5 Dilute the metabisulfite solution five-fold and wash the gel twice for 10 min.

6 Incubate in the dark for 1 h with Schiff's reagent.

7 Replace Schiff's reagent with metabisulfite solution (full strength) for 1 h in the dark.

8 Destain through four changes of destaining solution in the dark for 30 min each.

9 Replace destaining solution with preserving solution.

③ If a high background is encountered, increase the number of washes in step 4. Also some lots of gel contain a contaminant which stains [10]. To check this, stain a blank gel.

Pause point

[1] The gel can be left in fixing solution without harm.

Protocol 17e. Staining for lipoproteins

Reagents

Nile Red solution – 10 μg/ml Nile Red in acetone ⚠
Saturated solution of Oil Red O in 60% (v/v) ethanol ⚠
Saturated solution of Oil Red O and Fat Red 7B in 60% (v/v) ethanol
30% (w/v) Sucrose in distilled water
Sudan Black B solution – 100 mg in 200 ml of 60% (v/v) ethanol ①

Equipment

Gel dryer (preferably microwave)
Rocking platform
Staining dish
UV Transilluminator ⚠

Procedure

Prestaining with Nile Red

1 Put 20 μl of Nile Red solution into a glass tube and evaporate to dryness.

2 Add 50 μl of lipoprotein sample (0.5 mg/ml protein), swirl then add 5 μl 30% (w/v) sucrose.

3 Load 11 μl on to the gel.

4 Following electrophoresis, the gel is illuminated with a UV transilluminator emitting at 302 nm to visualize the proteins.②

Post-staining with Oil Red or Sudan Black

1 Cast the gel on to Gelbond for agarose (see *Protocol 5d*).

2 Following electrophoresis, dry the gel down using microwaves according to ref. 71.③

3 At room temperature, incubate the gel with shaking for 1 h in Sudan Black or 2 h in Oil Red.

4 After removing the gels from the staining bath, it will be necessary to wipe the stain sediment from the surface of the gel and from the Gelbond backing sheet.

Notes

This procedure will take approximately 2 hours.

① Filter immediately before use.

② The electrophoresis procedure must be completed within 2 h of prestaining before dissociation of the stain and lipid becomes appreciable. As such the use of minigels is recommended.

③ When agarose gels are dried rapidly using microwaves, the proteins become fixed into the matrix of the gel and subsequent diffusion is minimized. Plastic-backed agarose gels dried in this manner will swell to a fraction of their original thickness when incubated in aqueous solution.

Protocol 18. **Colloidal Coomassie Blue staining**

Reagents

Coomassie Blue G-250
Fixing solution – 500 ml absolute ethanol, 20 ml concentrated (83%) H_3PO_4, made up to 1 l with distilled water ⚠
Staining solution – 340 ml absolute methanol, 30 ml concentrated (83%) H_3PO_4 ⚠, 170 g ammonium sulfate, 460 ml distilled water

Equipment

Light box
Rocking platform or reciprocal shaker
Staining dish

Procedure

1 For gels about 1.5 mm thick, fix the gel in 150 ml of fixing solution for 2 h with rocking. See pause point 1 of *Protocol 17a.* ①

2 Replace with 150 ml of staining solution and incubate for at least 1 h with rocking.

3 Add approx. 100 mg of dry Coomassie Blue powder directly to the staining solution. ②

4 Rock the gels gently for at least 72 h. Complete partitioning can take as long as a week.

5 Following staining, wash the gels individually with water to remove any colloidal stain which may have settled on the gel surface.

Notes

This procedure will take approximately 75 hours.

① For gels of a different thickness, adjust the fixing and replacement time only; staining time is not influenced by thickness. Up to 10 gels can be stained in a single bath using this procedure. Multiply the volumes by 10.

② Coomassie Blue G 250 can be used directly. Purity of some lots is poor, thus further purification may be required. To achieve this, precipitate the stain from ammonium sulfate solution. Dissolve 4 g of dye in 250 ml of 7.5% (v/v) acetic acid and heat to 70°C. Stir in 75 g of solid ammonium sulfate and cool to room temperature. Collect the pellet by centrifugation. Air dry on a filter paper. Even stain of 95% purity will leave a green contaminant on precipitation.

Protocol 19. Silver staining of gels

Reagents

Absolute ethanol ⚠
Concentrated NH_4OH, 14.8 M
Developing solution – 0.15 ml of concentrated (37%) formaldehyde, 1.5 ml freshly prepared 1% (w/v) citric acid in distilled water, made up to 300 ml with 20% (v/v) ethanol ①
Ethanol solution – 20% (v/v) in distilled water
Fixing solution – 20% (v/v) absolute ethanol, 5.0% (v/v) glacial acetic acid, 2.5% (w/v) sulfosalicylic acid ⚠
Glacial acetic acid ⚠
Silver nitrate solution (freshly made) – 0.8 g $AgNO_3$ in 6 ml distilled water ②
Sodium hydroxide solution – 0.36% (w/v) NaOH. ⚠ (Vacuum degas if not fresh)

Equipment

Light box
Rocking platform or reciprocating shaker
Staining dish

Procedure

1 Fix the gel by incubating in 200 ml of fixing solution for at least 12 h with one solution change.③ ④ [1]

2 Wash the gel in 20% ethanol three times for 20 min each.

3 Mix 31.5 ml of NaOH solution with 2.1 ml of concentrated NH_4OH. Slowly add the 6 ml of $AgNO_3$ solution. Bring the volume to 90 ml with distilled water. Add 30 ml absolute ethanol. Bring the volume to 150 ml with distilled water.

4 Incubate the gel in the solution of step 3 for 1 h (maximum 1 h, 15 min)

Notes

This procedure will take approximately 16 hours.

① Prepare the developing solution just prior to use, during the third wash of step 6.

② The silver nitrate solution should be prepared on the day of use and kept in the dark. All others can be prepared in advance.

③ Volumes given are for individual gels, approximately 12 cm × 14cm × 1.5 mm. When staining more than one gel, use separate dishes for each so the slabs do not cover each other during rocking (particularly important during with rocking.

5 Decant and wash gels three times in water, 10 min each.

6 Decant and wash three times in 300 ml of 20% ethanol, 20 min each. If staining more than one gel at a time, space 1 min between gels.

7 Decant ethanol and pour on developing solution (space 1 min between gels). Developing time can vary between 10 and 20 min because the rate reaction varies for different proteins [7]; firm guidelines for developing time cannot be given.

8 If gels are developing too fast, when correct density is almost achieved, pour on 300 ml of 20% ethanol to slow down the reaction.⑤

9 When correct density is achieved, stop the reaction by adding 3 ml of glacial acetic acid. Wait 5 min then add 6 ml of concentrated (14.8 M) NH_4OH. Incubate for another 10 min. ⑥

10 Rinse gels four times in water, 10 min each, then store in 300 ml of 20% ethanol.⑦

step 7). The procedure is conveniently performed with five gels at a time.

④ Avoid fixing solutions containing trichloroacetic acid when gels are silver-stained by this procedure.

⑤ Many procedures use a development time of less than 5 min; gel-to-gel variation is difficult to manage. The procedure described here deliberately uses a slower development time (step 7) which makes gel to gel reproducibility more manageable.

⑥ With silver staining the image will keep developing until the reaction is terminated. Pay close attention to ensure that the pattern is not 'overdeveloped'.

⑦ Some formulations of silver stain fade with time. The gels should be evaluated, digitized or photographed within a day.

Pause point

[1] The gel can remain in fixing solution without harm.

Protocol 20. **Radiofluorography**

Reagents

5% (v/v) Glycerol in distilled water
20% (w/w) or 22.2% (w/v) PPO ⚠ (Amersham, Baker or Beckman) in DMSO ⚠ (full name, dimethylsulfoxide)
1 M Sodium salicylate, adjust to pH 5.7

Equipment

Film cassette with or without intensifying screens
Filter paper
Freezer, −70°C
Gel dryer
Rocking platform or wrist shaker
Staining dishes
X-ray film

Procedure 1

To impregnate a gel with PPO

1 Fix the protein pattern into the gel according to *Protocol 17*, and stain it if desired.

2 For a gel 1.5 mm thick, put the gel in destaining solution (*Protocol 17*) for at least an hour, regardless of whether you have stained it. Thicker gels and those cast on to backing sheets require longer times. If a stained gel is to be photographed, do so before step 3.

3 Remove all the water from the gel. Incubate it, with gentle rocking, for at least 30 min, at room temperature under a fume hood in at least four

Notes

This procedure will take approximately 72 hours.

① Removing water from the gel with DMSO causes it to shrink to a size that depends on %T; gradient gels become trapezoidal. Dehydration is complete when shrinkage stops.

② **Danger**: Wear gloves. DMSO is absorbed through the skin and can cause noxious halitosis that persists for days. PPO is carcinogenic.

③ An alternative to the DMSO procedure replaces water in the gel with increasing concentrations of acetic acid (25%, 50% then glacial). Finally, the gel is incubated in 20% (w/v) PPO in glacial acetic acid. PPO is precipitated and the gel dried as above before exposure [28].

volumes of DMSO. Repeat in a second bath of fresh DMSO for another 30 min. ① ② ③

4 Immerse the gel in PPO-DMSO solution (same volume as step 3) and incubate at room temperature for 3 h. ④

5 Rehydrate the gel by soaking it in 20 volumes of glycerol solution for 1 h at room temperature.

6 Dry the gel (*Protocol 27*).

7 Place the dried gel against a sheet of Kodak XAR film (or equivalent) and expose in a light-tight container at −70°C. ⑤

8 Following exposure, remove the film from the deep freeze. To keep the film free from condensation, remove it from the cassette before it warms up. Develop the film according to manufacturer's instructions.

To impregnate a gel with salicylate

1 Fix the gel and stain, if desired according to *Protocol 17*.

2 Soak the gel in 20 volumes of water for 30 min to avoid subsequent precipitation of salicylate by the low pH of the fixing and staining solutions.

3 Soak the gel in 10 volumes of salicylate solution for 1–2 h but not longer.

4 Perform steps 6–8 of the PPO procedure above. ⑤

④ DMSO can be recycled; reuse the second bath as well as the first. Excess PPO can be recovered from DMSO and reused. Add an equal volume of 15% ethanol in water and stir at room temperature. After the precipitate forms, add a large excess of water, filter the precipitate and allow to air dry.

⑤ Kodak XAR (or its equivalent from other manufacturers) is a good general purpose, double emulsion film which can be used for all emitters, for direct autoradiography, fluorography with impregnation and fluorography with intensifying screens. A less expensive general purpose film (not used with intensifying screens) is Kodak XLS. The trade-off is longer exposure time. Other films give slightly better results for specific applications. (i) A slightly sharper image can be obtained for direct autoradiography and impregnation fluorography if a single emulsion film (Kodak SB) is used. (ii) For direct autoradiography of ^{14}C and ^{35}S, Kodak Biomax MR is recommended. (iii) For ^{32}P and ^{125}I with intensifying screens, Kodak Biomax MS will require a shorter exposure time. Kodak SB is slightly more sensitive than Kodak XAR film to the salicylate fluor.

Pause point

[1] These procedures can be interrupted at any time. Prolonged incubation at any of the steps (with the exception of salicylate procedure, step 3) has no adverse effects.

VI IDENTIFICATION OF PROTEINS IN GELS

Introduction

It is frequently the case that demonstration of proteins in the gel by the techniques given in Chapter V is not sufficient and other identification procedures are required. The ultimate means of identification of proteins is amino acid sequencing. This is the most costly and, therefore, the least frequently used approach. The two major strategies used to assign identity to individual protein bands or spots are immunochemical techniques and enzymological techniques. Some applications combine the two. A third, less frequently used strategy is that of ligand (e.g. lectins and nucleic acids) binding. There are many variations in each category. For each approach, the identification can be attempted directly in the gel or by removing the protein from the gel. One method, immunosubtraction, removes the protein before electrophoresis.

Elution and renaturation

Three categories of schemes have been devised for elution of proteins from gels [1,2]: mechanical, diffusional and electrical. In the mechanical category, gels have been minced, crushed, homogenized and dissolved – agarose gels with low gelling temperatures have been melted. Diffusion alone where the band or spot is merely excised and incubated overnight in a buffer, is the least efficient method and is invariably combined with mechanical disruption of the gel. To accomplish this, cut out the area of gel of interest, suspend the gel slice in three times its original volume of the electrophoretic running buffer and homogenize in a glass/Teflon (Potter-Elvejhem) tissue grinder [1,3]. Place the suspension in a capped, siliconized tube and incubate at 4°C overnight with end-over-end rotation. Sediment the gel fragments by centrifugation and, after removal of the supernatant, resuspend the fragments in the same volume of fresh buffer and re-extract. Pool the eluates and concentrate, if necessary. Arrangements for the third category include electroelution into dialysis bags [4–8], sucrose gradients [2], hydroxylapatite [9], conductivity gradients [10], and capillary matrices [11]. Steady-state stacking [6] is also used. SDS

(0.1%) is sometimes recommended to keep the proteins in solution and to ensure that they migrate anodally. A few examples of applications which do not use SDS are, *N*-ethylmorpholine:acetic acid [7], steady-state stacking [6] and capillary matrices [11]. The most convenient procedure for electroelution, in the author's opinion, uses a simple device called an 'Elutrap' (Schleicher and Schuell) [12].

Proteins to be renatured must potentially recover from the untoward effects of detergents/reducing agents/polar solvents and acid/alcohol/high salt conditions. Enzymatic activity in the presence of SDS is rare but not unknown. Treatment with nonionic detergents is generally considered to be milder than SDS but variable. Some enzymes which are inactivated by Triton X-100 are active in Tween 20 [14]. A good discussion of the effect of nonionic detergents on immunoblotting has been published [15]. Some applications merely rely on soaking the SDS gel or blotting membrane for a short time in a buffer without detergents. This is a variable procedure, because proteins can differ in their rates of spontaneous refolding from fractions of seconds to days [16]. Other renaturation procedures rely solely on the slow process of diffusional elution.

Given that renaturation in the gel by rapid (and perhaps incomplete) buffer exchange may not be entirely satisfactory, alternative methods have been reported. A general method was proposed [17] which elutes the protein and uses 6 M urea to displace the SDS. SDS in solution is then adsorbed to Dowex 1 ion exchange resin, the sample is diluted slowly to reduce the urea to 0.6 M, and residual urea is removed by equilibrium dialysis. Other authors advocate acetone precipitation of the proteins to ensure complete removal of SDS [1]. The precipitates are redissolved in 6 M guanidine-HCl, then diluted. Other seemingly unlikely treatments have been reported. One is the addition of 1 mM mercaptoethanol [18]. Another is the inclusion of zwitterionic detergents in the SDS separating gel and the transfer buffer of the subsequent electroblot [19]. Finally, it is sometimes necessary to elute proteins from blotting membranes [20]. Here the elution required 2% SDS and/or 1% Triton X-100.

Recovery of activity by any of the above approaches appears to vary from protein to protein, suggesting that no universal method of renaturation is appropriate. Hence, comparison of enzyme activity or ligand/antibody binding in two bands of the same gel or in different gels requires caution since experimental observations can reflect variable renaturation as well as different amounts of protein present.

Methods available

Immunochemical identification

Western blotting (see *Protocol 21*)

Western blotting is the most popular method for identifying protein bands or spots using antibodies. It is the subject of complete textbooks (see refs 7 and 9 in Chapter I). The strategy for performing western blotting is to separate the proteins in a slab gel, transfer the pattern to a solid-phase membrane matrix where it will be immobilized, then react the membrane with specific antibodies. Following this, the location of the bound antibodies is indicated using 'reporters'. Reporters are labeled molecules, usually proteins, which can be radioactive, fluorescent, or linked to enzymes and are thus detectable in small amounts. Enzyme-linked reporters cleave soluble substrates whose products precipitate at the position of the bound enzyme. As an alternative to precipitable products, the enzymes can use chemiluminescence [21]. A review and discussion of the different reporters has been published [22]. In the most common practice, the target proteins are allowed to react with specific antibody (called the primary antibody). Its binding is located by a secondary antibody which carries the reporter and whose specificity is to the constant region of the primary antibody.

The blotting matrices commonly in use are nitrocellulose, fiberglass, nylon and polyvinylidene difluoride (PVDF) membranes. These

References

1. Hagar, D.A. and Burgess, R.R. (1980) *Analyt. Biochem.* **109**, 76–86.
2. Karsnas, P. and Roos, R. (1977) *Analyt. Biochem.* **77**, 168–175.
3. Djondjurov, L. and Holzer, H. (1979) *Analyt. Biochem.* **94**, 274–277.
4. Lewis, U.J. and Clark, M.O. (1963) *Analyt. Biochem.* **6**, 303-315.
5 Stephens, R.E. (1975) *Analyt. Biochem.* **65**, 369–379.
6. Nguyen, N.Y., DiFonzo, J. and Chrambach, A. (1980) *Analyt. Biochem.* **106**, 78–91.
7. Bhown, A.G., Mole, J.E., Hunter, F. and Bennett, J.C. (1980) *Analyt. Biochem.* **103**, 184–190.
8. Tuszynski, G.P., Damsky, C.G., Fuhrer, J.P. and Warren, L. (1977) *Analyt. Biochem.* **83**, 119–129.
9. Ziola, B.R. and Scraba, D.G. (1976) *Analyt. Biochem.* **72**, 366–371.
10. Stralfors, P. and Belfrage, P. (1983) *Analyt. Biochem.* **127**, 7–10.
11. Thelu, J. (1988) *Analyt. Biochem.* **172**, 124-129.
12. Jacobs, E. and Clad, A. (1986) *Analyt. Biochem.* **154**, 583–589.
13. Mozhaev, V.V., Berezin, I.V. and Martinek, K. (1987) *Meth. Enzymol.* **135**, 586–596.
14. Heegaard, P.M.H. (1988) in *Handbook of Immunoblotting of Proteins*, Vol. 1 (O.J. Bjerrum and N.H.H. Heegaard, eds), pp. 221–225. CRC Press, Boca Raton, FL.

membranes are available under a variety of different trade-names and each has advantages and disadvantages. A thorough description of the binding, background and stripping characteristics of these membranes is available [23]. Briefly, nitrocellulose is least expensive, but it is brittle when dry. Fabric-reinforced nitrocellulose is available, but it is more expensive. Nylon membranes have a higher protein binding capacity per unit area than nitrocellulose, but the increased capacity can lead to higher backgrounds. Nylon membranes can be stripped and reprobed many times whereas nitrocellulose can be stripped only once or twice. PVDF membranes are the most expensive but they are more easily stripped and reprobed. They are also the membrane of choice for amino acid microsequencing.

Transfer of the proteins from the gel to the membrane can be effected by capillary action or using an electric field. Capillary blotting is an overnight procedure [24], but can be hastened by using a vacuum table (an ordinary vacuum gel dryer) [25], or a dedicated instrument (for example, Hoefer Trans-vac). Both surfaces of the gel must be open; gels can be removed from plastic backing sheets using a slicing instrument, similar to a wire cheese cutter (Pharmacia). As an alternative to blotting procedures, detection of proteins using antibodies without transferring the pattern to a membrane matrix is possible. The problems of diffusional loss of resolution of the pattern during the lengthy incubation and washing steps required have been

15. Bjerrum, O.J., Selmer, J.C. and Lihme, A. (1987) *Electrophoresis*, **8**, 388–397.
16. Anfinsen, C.B. (1986) in *Protein Engineering* (M. Inouye and R. Sarma, eds), pp. 3–13. Academic Press, New York.
17. Weber, K. and Kuter, J. (1971) *J. Biol. Chem.* **246**, 4504–4509.
18. Chang, L.M.S., Plevani, P. and Bollum, F.J. (1982) *Proc. Natl Acad. Sci. USA*, **79**, 758–761.
19. Fulop, M.J., Webber, T. and Manchee, R.J. (1992) *Analyt. Biochem.* **203**, 141–145.
20. Szewczyk, B. and Summers, D.F. (1988) *Analyt. Biochem.* **168**, 48–53.
21. Pringle, M.J. (1993) *Adv. Clin. Chem.* **30**, 89–183.
22. Van Oss, C.J. and Van Regenmortel, M.H. (1994) *Immunochemistry*, pp. 925–936. Marcel Dekker, New York.
23. Tovey, E.R. and Baldo, B.A. (1989) *J. Biochem. Biophys. Meth.* **19**, 169–184.
24. Renart J., Reiser, J. and Stark, G.R. (1979) *Proc. Natl Acad. Sci. USA*, **76**, 3116–3120.
25. Peferoen M., Huybrechts, R. and DeLoff, A. (1979) *FEBS Lett.* **145**, 369–372.
26. Shainoff, J. (1993) *Adv. Electrophor.* **6**, 64–177.
27. Towbin, H., Staehlin, T. and Gordon, J. (1979) *Proc. Natl Acad. Sci. USA*, **76**, 4350–4354.
28. Montelaro, R.C. (1987) *Electrophoresis*, **8**, 432–438.
29. Baillod, P., Affolter, B., Kurt, G. and Pflugshaupt, R. (1992) *Thromb. Res.* **66**, 745–755.

solved by the recent introduction of glyoxyl agarose as a gel electrophoresis medium [26].

Transfer can also be driven by electrophoresis [27]. This is more rapid than capillary blotting and is reported to transfer proteins efficiently. Caution must be exercised to ensure that the proteins move toward the membrane. If blotting is from an SDS gel, the proteins will already be negatively charged, but not so if blotting is from an isoelectric focusing gel or some other system. As a precaution against protein loss, include SDS in the transfer buffer or use one membrane on each side of the gel. There are two approaches to electroblotting. The original, more common method, uses a buffer-filled tank in which the gel and blotting membrane are immersed vertically. The other method is 'semi-dry' electroblotting in which the gel/membrane sandwich is placed horizontally between buffer-saturated filter papers. The entire stack then is placed between graphite electrode slabs whose areas are each larger than the stack. The entire surface of the slabs is conductive. Semi-dry blotting is faster but the instruments are more expensive.

Problems: Western blotting has serious limitations. Sample preparation, electrophoresis and blotting conditions can destroy the relevant epitopes. Renaturation is often inefficient and/or incomplete. Because specific antibodies are most frequently raised to proteins in their native conformations, the binding of the antibodies to proteins

30. Altland, K. and Hackler, R. (1981) *Electrophoresis*, **2**, 49–54.
31. Boyle, M.D.P. and Metzger, D.W. (1994) in *Antibody Techniques* (V.S. Malik and E.P. Lillehoj, eds), pp. 177–209. Academic Press, San Diego.
32. Gersten, D.M. and Zapolski, E.J. (1991) *Adv. Electrophor.* **4**, 50–79.
33. Kumar, A., Kim, H.R., Sobol, R.W., Becerra, S.P., Lee, B.J., Hatfield, D.L., Suhadolnik, R.J. and Wilson, S.H. (1993) *Biochemistry*, **32**, 7466–7474.
34. Schiff, L.A., Nibert, M.L. Co, M.S., Brown, E.G. and Fields, B.N. (1988) *Mol. Cell Biol.* **8**, 273–283.
35. Gabriel, O., and Gersten, D.M. (1992) *Analyt. Biochem.* **203**, 1–21.
36. Ohlsson, B.G., Westrom, B.R. and Karlson, B.W. (1987) *Electrophoresis*, **8**, 415–420.
37. Gersten, D.M. and Gabriel, O. (1992) *Analyt. Biochem.* **203**, 181–186.
38. Mochenko, G.P. (1994) *Handbook of Detection of Enzymes on Electrophoretic Gels.* CRC Press, Boca Raton, FL.
39. Patel, D. (1994) *Gel Electrophoresis: Essential Data.* John Wiley & Sons Ltd, Chichester.
40. Lantz, M.S. and Ciborowski, P. (1994) *Meth. Enzymol.* **235**, 563–594.
41. Kinzkofer-Peresch, A., Petestos, N.P., Fauth, M., Kogel, F., Zok, R. and Radola, B.J. (1988) *Electrophoresis*, **9**, 497–501.

treated as described previously is unpredictable.

Immunosubtraction (see *Protocol 22*)

Immunosubtraction sidesteps epitope denaturation by performing the antibody–antigen interaction before electrophoresis. As such, it is the preferred technique for many applications. The strategy in performing an immunosubtraction experiment is to divide the sample in two and react one aliquot with specific antibodies and the other with control antibodies. Next the antibodies along with their bound antigen are removed from the sample. Both aliquots can then be treated with SDS or other denaturants and are subsequently electrophoresed in adjacent lanes (or companion gels for two-dimensional separations). Comparison of the two lanes identifies the band(s), subtracted by the antibody. The specific antibody with its attached antigen can be immunoprecipitated with a second antibody. (Immunoprecipitates can also be carried out using primary antibody only, if the antigen and antibody are near the equivalence point.) Guidelines on immunosubtraction of human serum proteins by primary antibody have been published [30]. Alternatively, staphylococcal Protein A or Protein G can be used [31]. It is convenient to attach either of these molecules to a beaded matrix, or to purchase them already conjugated. The use of Protein A–Sepharose (Pharmacia) or Protein G–Agarose (Pierce) is easier than immunoprecipitation using a second antibody.

42. Christensen, J. and Houen, G. (1992) *Electrophoresis*, **13**, 179–183.
43. Patton, W.F., Lam, L., Su, Q., Lui, M., Erdjument-Bromage, H. and Tempst, P. (1994) *Analyt. Biochem.* **220**, 324–335.
44. Aepinus, C., Voll, R., Broker, M. and Fleckenstein, B. (1990) *FEBS Lett.* **263**, 217–221.
45. Gersten, D.M. and Marchalonis, J.J. (1978) *J. Immunol. Meth.* **24**, 305–309.
46. McGrew, B.R. and Green, D.M. (1990) *Analyt. Biochem.* **189**, 68–74.
47. Kuhnl, P., Schmidtmann, U. and Spielmann, W. (1977) *Hum. Genet.* **35**, 219.

Protocols provided

21a. *Capillary blotting from a polyacrylamide slab gel to nitrocellulose*
21b. *Vertical electroblotting to nitrocellulose, nylon-66 or PVDF*
21c. *Northwestern blotting*
22. *Immunosubtraction*
23. *Infusion of reagents into gels: demonstration of β-galactosidase*
24a. *One-step zymogram using an indicator matrix – demonstration of DNase*
24b. *Coupled zymogram using an indicator matrix – demonstration of phosphoglucomutase*
25. *Activity gels*

A variation on this procedure is to pre-incubate the primary antibody with Protein A–Sepharose or Protein G–Agarose and wash the beads prior to incubation with the sample. In this way, contamination of the sample with excess antibodies is avoided.

Associated methodologies

Associated technology to the basic western transfer has also been developed. Among the most important innovations is the use of membrane-bound proteins for Cleveland-style peptide mapping and amino acid microsequencing. The ultimate means of identifying an unknown spot or band is by its N-terminal sequence. Other important, related procedures are northwestern and southwestern blotting (see *Protocol 21*), in which nucleic acid-binding proteins are identified and RNA– or DNA–protein interactions can be assessed. Southwestern and northwestern procedures seek to identify DNA- and RNA-binding proteins, respectively, on protein blots. To perform a northwestern or southwestern experiment, first separate the protein mixture in a slab gel and transfer the pattern to a membrane using a standard western technique. Next, the membrane is incubated with a radioactive DNA or RNA probe, usually having a ^{32}P-label [32]. The location of the bound radioactivity is determined by autoradiography or fluorography. Examples where northwestern and southwestern blotting have been combined with other gel techniques include peptide mapping [33] and microsequencing from blots or metal binding [34].

Zymographic identification (see *Protocols 23–25*)

Assignation of enzymatic activity to a particular band or spot is one of the most powerful (and tricky) means of identification. In addition to demonstrating activity, one must also be convinced that the absence of demonstrable activity indeed signifies absence of the enzyme, not inhibition or inactivation. Often the conditions adopted represent a compromise between optimal separation and optimal enzymology. Considerations unique to zymography include the potential untoward effects of contaminants in the gel (see *Protocol 6*), detergents, inhibitory effects of tracking dyes, diffusability of reagents into the gel, stability/renaturability of the enzyme.

The five approaches for demonstrating enzymatic activity in gels [35] are: (i) simultaneous capture of substrate product, (ii) postincubation of substrate product, (iii) autochromic methods, (iv) indicator-matrices, and (v) copolymerization of substrate into separating gel (also called 'activity gels'). Once a procedure is selected, it is helpful to perform a pilot experiment. Cast a small gel, of the type to be used, on to a silanized microscope slide or plastic backing sheet. Make a small well and allow the active enzyme to diffuse into the gel. Perform the zymographic procedure without electrophoresis. This will indicate whether enzymatic activity is inhibited by the gel conditions and whether enough enzyme can be loaded to achieve detectable substrate conversion. In all approaches except activity gels, most of the substrate conversion takes place at the gel surface. Furthermore, when enzymes must be renatured, the process is rarely complete. Consequently, substrate conversion is usually catalyzed by only a fraction of the amount of enzyme actually present in a band or spot.

The number of zymographic techniques is enormous and they are necessarily tailored to individual enzymes. This chapter gives specific examples of three of the most commonly used approaches, infusion of substrate into the electrophoresis gel, overlay matrix and activity gel. Additional advice on specific enzymes has been published [35, 37–39].

Infusion of substrate (see *Protocol 23*)
The simplest zymographic technique is an autochromic method where the cleavage product is colored and insoluble. The example given here identifies β-galactosidase (E.C. 3.2.1.23), separated in an SDS gel and renatured by one soaking step. β-galactosidase cleaves the substrate X-gal (full name is 5-bromo, 4-chloro, 3-indoyl-β-D-galactopyranoside) to yield a blue product.

Overlay matrices (see *Protocol 24*)
The advantages of overlay indicator matrices are:

(i) Substrates which do not have insoluble cleavage products can be brought into contact with the separating gel. When substrate conversion is rapid, the overlay can be read before diffusion is appreciable.

(ii) An overlay can be impregnated with all the reagents necessary to demonstrate coupled reactions, obviating sequential incubation of the separating gel in different solutions.

(iii) More than one indicator matrix can be used on the same slab gel so that a single separation is not limited to one zymogram.

The strategy is to prepare an overlay of filter paper, membrane or wide-pore agarose gel impregnated with the necessary substrates and cofactors. Following equilibration of the separating gel to the pH optimum of the enzyme, the overlay matrix is abutted against the separating gel and the reaction occurs at the interface of the separating gel and the overlay. A protein blot can be used in place of the separating gel.

Single step reaction (see *Protocol 24a*): For the demonstration of DNase (E.C. 3.1.x.x), an agarose indicator matrix containing high molecular weight DNA and ethidium bromide is used. High molecular weight DNA will intercalate ethidium bromide and fluoresce orange when illuminated at 312 nm. Areas of the indicator gel in which the substrate has been degraded will not fluoresce and appear as dark bands on a bright orange background.

Coupled reaction (see *Protocol 24b*): The demonstration of phosphoglucomutase (E.C. 2.7.5.1), which catalyzes the conversion of glucose-1-phosphate to glucose-6-phosphate in the presence of glucose-1,6-bisphosphate, is a good example of a coupled reaction. The indicator matrix contains the required substrates and cofactors plus glucose-6-phosphate dehydrogenase. The target enzyme, phosphoglucomutase, generates its product, glucose-6-phosphate, which is immediately degraded by glucose-6-phosphate dehydrogenase. The action of glucose-6-phosphate dehydrogenase, in turn consumes $NADP^+$ to yield NADPH. NADPH is detected when its reducing equivalent is transferred to Thyazol Blue (also called MTT or [3-(4,5-dimethylthiazol-2-yl)-2,5-diphenyltetrazolium bromide]) by mediation of PMS (phenazine methosulfate). This results in a blue color.

Activity gels (see *Protocol 25*)
A polyacrylamide separating gel with pores of about 7%T, 2.6%C or smaller will entrap high molecular weight proteins and nucleic acids during gel polymerization. The requirement for macromolecules limits the use of activity gels to demonstrate proteolytic enzymes [40], nucleases, polymerases or group transferases [37] where the precursors and products differ vastly in size. Acid solubility of the smaller molecules and insolubility of the macromolecules is exploited. The strategy for performing a zymography experiment for proteolytic

enzymes in an activity gel is to polymerize the slab gel (both SDS and nondenaturing gels are used) to contain a high molecular weight protein substrate. Following electrophoresis, the gel is equilibrated in buffer containing any necessary cofactors, at the pH optimum of the enzyme, and digestion proceeds. The gel is then fixed in acid/alcohol and stained. Areas of the gel where entrapped protein has been enzymatically degraded will appear as clear bands against a stained background.

Choice of methods

The choice of identification methods is based, of course, on whether the protein of interest has a known enzymatic activity, known ligand-binding activity or whether specific antibodies are available. The decision to use *in situ* methods or to remove the protein from the gel involves a compromise; methods which take place in the gel require less time and fewer steps but sensitivity may be lower and background higher. Moreover, the most common gel techniques (e.g. SDS) denature the protein and removal from the gel may be necessary for renaturation.

When the binding or enzymatic activity is destroyed, three options are available: (i) perform the identification interaction prior to the electrophoresis procedure (see *Protocol 22*); (ii) renature the protein in the gel prior to identification procedures; or (iii) remove the protein from the gel by blot-transfer or elution, then renature and perform the identification procedure with the protein in solution or anchored to a blotting membrane.

Western blotting is the most popular method for immunochemical identification. Each of the blotting membranes and transfer methods has its advantages and disadvantages. Some of the considerations are as follows:

(i) Not all proteins will bind to all matrices. Preliminary experiments are required in which the crude mixture is spotted, fixed to the matrix and tested for retention using the intended identification technique. Alternatively, some matrices will bind everything, thereby causing a high background. The unoccupied binding sites of such supports can be blocked following transfer by incubation in a solution of proteins which will block non-specific sites (e.g. BSA, soluble fish gelatin, powdered nonfat milk, Irish cream liqueur).

(ii) Some electroblotting procedures require transfer in the presence of denaturants such as acetic acid [27], methanol [28], or SDS (see *Protocol 21b*).

(iii) Electroblotting and capillary blotting require that both sides of the slab gel be available. Acceptable transfer from gels cast on backing sheets can only be achieved by vacuum blotting if the gel is removed from its backing.

(iv) Overnight transfer for capillary blots may be too long to maintain the activity of some enzymes.

(v) Larger proteins are frequently transferred to membranes less efficiently than smaller ones. The use of SDS–agarose electrophoresis followed by blotting from that larger-pored matrix has been recommended [29] or the use of glyoxyl agarose may be indicated [26]. Generally, protein is eluted more efficiently from the gel by electroblotting than by capillary blotting.

Capillary blotting to nitrocellulose is the commonest, easiest, least expensive blotting procedure and serves most applications. It does not require equipment and does not use denaturing electrolytes. As such, many workers try this method first.

Zymography is chosen over immunochemical identification when specific antibodies are unavailable or to sidestep the issue of antibody cross reactivity. When there is insufficient enzyme activity at the gel surface, consider performing zymography on a blot. Blotting serves to concentrate the protein from the gel slab on to an essentially two-dimensional matrix. If this strategy, by itself, does not give enough active enzyme for rapid substrate conversion, the immobilized enzyme can be incubated with substrate for very long periods of time without detachment from the membrane or diffusion of the pattern. Another advantage of blotting [36] is that the blotting matrix can be incubated in more than one substrate. Elution of the enzyme should be used as a last resort.

Protocol 21a. Capillary blotting from a polyacrylamide slab gel to nitrocellulose

Reagents

10% (v/v) Glacial acetic acid ⚠ in distilled water
Blocking buffer – TBS-Tween containing 7% (w/v) nonfat dry milk
DAB solution – 1 mg 3,3′-diaminobenzidine ⚠ (Dako or Sigma) per 40 ml TBS. Make 30 min before use and keep in the dark. Immediately before use, add H_2O_2 ⚠ to a final concentration of 0.01% (v/v)
Peroxidase-conjugated second antibody (Dako or equivalent)
Primary antibodies
Transfer buffer – 0.02 M phosphate buffer, pH 8.0
TBS-Tween – 10 mM Tris-HCl, pH 7.2, 0.145 M NaCl, 0.1% (v/v) Tween 20

Equipment

Filter paper
Glass dish
Glass plates
Nitrocellulose blotting membrane (Immobilon-NC, Millipore or equivalent)
Paper towels
Plastic wrap or Parafilm
Rocking platform

Procedure ①

1 Pre-wet an Immobilon-NC membrane (or equivalent) by floating it in a dish containing transfer buffer. Do not immerse the membrane since this may cause air to become trapped inside the pores.

2 Pre-wet six sheets of Whatman No.1 filter paper in transfer buffer.

3 Fill a square or rectangular glass dish with transfer buffer, and cover the

Notes

This procedure will take approximately 21 hours.

① Immunodetection will detect proteins transferred from fingerprints. Wear disposable plastic gloves.

② Sometimes the weight used squeezes protein down into the wick instead of drawing it up into the membrane. If so, blotting can be made unidirectional by interposing a dish with a glass plate which extends over two of the four sides. Drape the six thicknesses of filter paper over the plate such that both ends are immersed as wicks in the buffer.

4 Position the slab gel on the filter paper, being careful to avoid air bubbles between the gel and the paper.

5 Use plastic wrap or Parafilm to 'frame' the gel. This is to ensure that the only fluid flow is through the gel and not around it. The entire surface of the filter paper except that occupied by the gel should be covered by the plastic.

6 Place the saturated membrane on the gel, again avoiding air bubbles.

7 Cover the membrane with two sheets of dry filter paper (S & S no. 597).

8 Prepare a mat of dry paper towels, 25 mm thick, and place these on top of the dry filter papers.

9 Cover the stack with a glass plate, and put a 1 kg weight on the glass.②

10 Incubate overnight. To assess the extent of protein transfer, stain the gel with Coomassie Blue (*Protocol 17a*). Alternatively, stain the membrane by incubation for 3 h in 0.1% (w/v) Naphthol Blue-black (also called Amido Black 10B and Buffalo Black, CI 20470) in distilled water. Destain by incubation in water.③④

11 Block the unoccupied membrane binding sites by incubating it in blocking buffer for 2 h at room temperature in a shallow dish on a rocking platform.

dialysis membrane between the filter paper wick and the gel [41].

③ An infrequent practice following transfer is to bake nitrocellulose membranes *in vacuo* at 80°C for 2 h to fix the proteins. A disadvantage is that this may affect subsequent antibody binding.

④ If a record of the stained membrane is required, photograph or photocopy it now (*Protocol 26*). Membranes to be used for amino acid microsequencing (PVDF membranes are recommended for this), should be stained with those that are fully dissociable from the membrane. Use Coomassie Blue according to [42] or an iron chelating dye [43].

⑤ Dilution for primary and secondary antibody and incubation time in antibody must be determined experimentally. For a commercial primary antiserum, start with 1:500, incubated for 1 h. For a commercial second antibody (e.g. peroxidase-conjugated goat anti-rabbit) start with 1:1000 for 1 h.

⑥ The brown precipitate may fade over long periods of time.

12 Dilute the primary antibody appropriately in TBS-Tween. Incubate the membrane for 1 h at room temperature or 4°C overnight, if necessary.⑤ [1]

13 Wash off the unbound primary antibody by incubating the membrane in TBS-Tween with rocking for 10 min. Repeat twice. Thorough washing at this stage is critical.

14 Dilute peroxidase-conjugated second antibody 1:1000 in TBS-Tween and incubate the membrane for 1 h at room temperature. ⑤

15 Repeat step 13 for six washes. Incomplete washing here is the primary cause of high background.

16 Immerse the membrane in DAB solution at room temperature with rocking for 10 min.

17 Stop the reaction by immersing the membrane in 10% (v/v) acetic acid.⑥

Pause point

[1] If the membrane is blocked adequately, increasing the incubation time with the primary antibody will not increase the background.

Protocol 21b. **Vertical electroblotting to nitrocellulose, nylon-66 or PVDF**

Reagents

10% (v/v) Glacial acetic acid ⚠ in distilled water

Blocking buffer – TBS-Tween containing 7% (w/v) nonfat dry milk

DAB solution – 1 mg 3,3′-diaminobenzidine⚠ (Dako or Sigma) per 40 ml TBS. Make 30 min before use and keep in the dark. Immediately before use, add H_2O_2 ⚠ to a final concentration of 0.01% (v/v)

Peroxidase-conjugated second antibody (Dako or equivalent)

Primary antibodies

TBS-Tween – 10 mM Tris-HCl, pH 7.2, 0.145 M NaCl, 0.1% (v/v) Tween 20

Transfer buffer solution – 20 mM Tris, 160 mM glycine, 20% (v/v) methanol, 17 mM NaCl, 0.1% (w/v) SDS. Adjust to pH 8.3 ①

Equipment

Blotting membrane (nitrocellulose, nylon or PVDF)

Electroblotting apparatus (BioRad Transblot or equivalent)

Filter paper

Glass dish

Magnetic stirrer

Power supply

Rocking platform

Procedure

1 Pre-wet the membrane by flotation in transfer buffer for nitrocellulose. For nylon and PVDF, wet with 20% methanol then immerse in transfer buffer (see step 1 of *Protocol 21a*).

2 Cut sheets of Whatman 3MM filter paper to the same size as the sponges (for thin gels) or fiber pads (thick gels) in the gel holder. Saturate the filter paper and sponges or pads with transfer buffer.

Notes

This procedure will take approximately 8 hours.

① This procedure uses a methanol- and SDS-containing transfer buffer, which may denature some proteins.

② Charged nylon-66 has a very high binding capacity and is useful for electroblotting but, due to its charge, it cannot be destained.

③ The time required for transfer depends on the gel

3 Equilibrate the gel in transfer buffer (20 min for a 0.75 mm thick gel, 30 min for a thicker gel).

4 Place one plastic grid (cathodal side) on to a glass plate.

5 Assemble the 'sandwich'. Expel any air bubbles trapped between the layers by rolling the surface with a test tube. Place a saturated sponge or fiber pad on the cathode grid. Cover it with a sheet of wet filter paper. Place the gel on top of the filter paper and the membrane sheet on top of the gel. Cover the membrane with a sheet of pre-wetted filter paper then another pad or sponge.

6 Place the anodal grid on top and while grasping the sandwich firmly, lock the cassette and insert into the tank, which has previously been half filled with transfer buffer. The proteins in the gel will migrate toward the anode. Make sure that the sandwich is oriented with the membrane between the gel and the anode.

7 Fill the tank with transfer buffer. It is important to use a full tank because the buffer is the coolant.

8 Run the system at 100 V, constant voltage, for 2 h. If transfer is incomplete, precool the buffer and run the system in the cold room at 500 V for 2 h. Circulate the buffer using a magnetic stirrer. To determine if transfer is adequate, stain the gel or stain the membrane according to note 4 of *Protocol 21a*. ② ③

9 See steps 11–17 of *Protocol 21a* for blocking and developing instructions.

thickness and protein size relative to gel pore sizes. For 12%T, 0.75 mm thick gels, transfer at 100 V for 2 h is usually sufficient to remove proteins of $M_r < 100\,000$.

Reagents

Binding buffer solution – 50 mM NaCl, 1 mM EDTA, 0.02% (w/v) bovine serum albumin, 0.02% (w/v) Ficoll 400, 0.02% (w/v) polyvinylpyrrolidone 40 ①

Escherichia coli tRNA (Sigma)

Equipment

Intensifying screen (Dupont lightning plus or equivalent)

Materials for western transfer (*Protocols 21a* or *21b*)

Perspex shielding

X-ray film (Kodak XAR or equivalent)

Procedure

1 Perform a western transfer according to the instructions given in *Protocol 21a*, steps 1–10 or *Protocol 21b*, steps 1–8.

2 Block the unoccupied binding sites by incubating the membrane at room temperature for 1 h in binding buffer containing 20 μg/ml *E. coli* tRNA [44]. If probing for tRNA binding proteins, substitute 10 μg/ml sheared salmon sperm or calf thymus DNA for tRNA.

3 Incubate the membrane for 10–60 min at room temperature in binding buffer containing 10^4–10^6 d.p.m./ml ^{32}P-RNA. ②

4 Wash the membrane three times in binding buffer without BSA, Ficoll and PVP at room temperature for 5–15 min. ③

5 Allow the membrane to air dry.

6 Detect the bound radioactivity by autoradiography. Expose the blot to Kodak XAR using an intensifying screen according to *Protocol 20.*

Notes

This procedure will take approximately 24–48 hours.

① Some authors double the concentration of BSA, Ficoll and PVP.

② Incubation time depends on both specific activity of the RNA and binding constant of the protein, and is necessarily determined experimentally. Likewise, washing time is variable.

③ **Caution**: When working with radioisotopes take all necessary precautions. Dispose of the first wash to the radioactive waste.

Protocol 22. Immunosubtraction

Reagents

Binding buffer solution – 0.1 M sodium acetate, pH 5.0
Elution buffer – 0.15 M NaCl, 0.1 M acetic acid, pH 2.9
Protein G-Agarose (Pierce)
Specific antiserum and control serum

Equipment

Benchtop microcentrifuge
Microcentrifuge tubes

Procedure

1 Equilibrate Protein G-Agarose as a 50% slurry in binding buffer.

2 Add 200 μl of slurry to each of three microcentrifuge tubes. Centrifuge at 200 *g* and replace the binding buffer with 160 μl of specific antiserum or 160 μl control serum or 160 μl buffer.①

3 Incubate in the same closed microcentrifuge tube overnight at 37°C, then wash the beads 20 times in binding buffer. [1]

4 Centrifuge the washed beads for 10 min at 200 *g* and replace the supernatant with sample.②

5 Incubate at 37°C overnight. If the sample is larger than about 200 μl, incubate with end-over-end rotation.

6 Centrifuge for 10 min at 200 *g*.

Notes

This procedure will take approximately 18 hours.

① When using Protein A or Protein G conjugated to beads, it is necessary to determine whether any of the proteins in the sample bind nonspecifically to Protein A, Protein G or the bead matrix. Thus an additional control of beads without serum is used.

② The amount of sample to be incubated with the antibody-coated beads must be determined experimentally. A good starting point is 500 μl of sample containing up to 1 mg total protein. When using conditioned culture medium or other dilute solution as the sample, as much as 2 ml of sample can be used with 100 μl of beads.

③ Protein loss can be minimized by compressing the beads with a high-speed centrifugation (10 000 *g*) in step 6.

④ When elution is used the antibody, if not covalently linked

7 Prepare the supernatant for the selected electrophoretic procedure. Measure the amount of protein in the supernatant because some will be lost to the internal volume of the beads. As an added control, the target antigen is sometimes eluted from the bead and electrophoresed in a separate lane. For elution, wash the beads in binding buffer following step 6 and incubate them in 200 μl of elution buffer (0.15 M NaCl, 0.1 M acetic acid, pH 2.9) for 10 min at room temperature [45]. ③ ④

8 Perform electrophoresis, fix and stain the gel, compare lanes. ⑤

to the bead, will also elute and will constitute the majority of protein.

⑤ For positive identification of bands using immunosubtraction, it is not necessary to subtract the entire band; a reduction in the amount of protein indicates recognition by the antibody.

Pause point

[1] Longer incubations are usually not detrimental.

Protocol 23. Infusion of reagents into gels – demonstration of β-galactosidase

Reagents

Reaction buffer – 0.1 M sodium phosphate buffer, pH 7.0,containing 10 mM KCl, 1 mM $MgSO_4$ and 39 mM 2-mercaptoethanol
Washing solution – 0.04 M Tris-HCl, pH 9.0, 0.02% NaN_3, 2 mM EDTA, 1% (w/v) casein
X-gal stock solution – 20 mg/ml X-gal in dimethylformamide. Store at −20°C ⚠

Equipment

Glass dish
Rocking platform

Procedure

1 To identify β-galactosidase, prepare an SDS slab gel according to *Protocol 12* such that each of the samples is run in duplicate. Following electrophoresis, cut the gel slab in half. Fix and stain half the gel to demonstrate all the proteins present (*Protocol 17a*) and use the other half for zymography.①

2 From the zymogram half remove SDS by incubating, at room temperature with rocking, for 2 h through four changes of washing solution, or washing solution without casein but containing 25% (v/v) isopropanol.② ③ ④

3 Equilibrate the gel through two washes, 30 min each, at room temperature in 10 volumes of reaction buffer.

Notes

This procedure will take approximately 4 hours.

① Use the highest quality SDS available. Some lots contain inhibitory contaminants.

② The incubation times are given for a gel 0.75 mm thick. For thicker gels, increase the times proportionally.

③ DE-52 (diethylaminoethyl cellulose, Whatman Bioproducts) can be substituted for casein [46]. Swell DE-52 resin in wash solution (without casein). Wash the resin until the wash has the same conductivity as fresh solution. Adjust the final volume to make a 10% slurry and incubate the gel with gentle rocking in 10 volumes of the 10% slurry for 2 h with four slurry changes. Casein and DE-52 presumably accelerate SDS removal by creating a

4 Incubate the gel at 37°C in 10 volumes of reaction buffer to which 1 ml of X-gal stock solution per 100 ml of reaction buffer has been added. Continue incubation until a blue color appears.

5 If no color appears following overnight incubation, it is unlikely that active enzyme is present.

6 Compare the stained half of the gel with the zymogram to identify which band contains the enzyme.

false concentration gradient. The mechanism by which aqueous isopropanol works is unclear.

④ Casein is sometimes contaminated with proteolytic enzymes – inactivate them by autoclaving the washing solution.

Protocol 24a. One-step zymogram using an indicator matrix – demonstration of DNase

Reagents

Agarose solution, 2% (w/v) in distilled water (Seakem GTG, FMC Bioproducts or equivalent) ①

Cacodylate solution – 0.1 M sodium cacodylate ⚠ at pH 6.5 containing 0.2 mM $CaCl_2$, 20 mM $MgCl_2$. Add ethidium bromide ⚠ to a final concentration of 50 μg/ml and salmon sperm DNA to a concentration of 0.5 mg/ml

Equipment

UV transilluminator ⚠

Procedure

1 Prepare an overlay indicator matrix, on a sheet of Gelbond backing, which has the same dimensions as the separating gel and is 1 mm thick or less. Calculate the required volume, then heat 50% of that volume of cacodylate solution to 60°C. Melt the agarose solution and allow to cool to 60°C. Mix equal amounts of cacodylate solution and agarose solution and pour the gel. ①

2 Allow the overlay matrix to air dry, or dry it at 50°C. [1]

3 Separate the protein mixture in an SDS gel or isoelectric focusing gel, according to the instructions in *Protocols 10,11* or *12*.

Notes

This procedure will take approximately 1 hour.

① See notes 3 and 4 of *Protocol 4* for melting agarose, and *Protocol 5d* for casting an agarose gel on to Gelbond backing.

② When trans-illuminating the 'sandwich', make sure the separating gel is next to the UV light. Unless Gelbond NF is used, the backing sheet will not permit the passage of UV light.

4 Overlay the dried indicator gel on the separating gel and incubate the 'sandwich' at 37°C in a humid chamber. If the indicator gel has curled, weigh it down with a glass plate to ensure intimate contact with the separating gel. Monitor the progress of the reaction by illumination at 312 nm. ②

Pause point

[1] For DNase, the overlay matrix can be prepared in advance and stored dry, in the dark, at room temperature for several weeks.

Protocol 24b. **Coupled zymogram using an indicator matrix – demonstration of phosphoglucomutase**

Reagents

Agarose, Isogel, Seakem Gold (FMC Bioproducts) or equivalent
Glucose-1,6-bisphosphate
Glucose-1-phosphate
Glucose-6-phosphate dehydrogenase
$NADP^+$
Phenazine methosulfate (PMS)
Thyazol Blue (MTT, [3-(4,5-dimethylthiazol-2-yl)-2,5-diphenyl tetrazolium bromide])
Tris buffer solution – 1.8 g Tris-HCl, 1.0 g $MgCl_2 \cdot 6H_2O$, 100 mg NaN_3 made up to 100 ml with distilled water. Adjust to pH 9.0

Equipment

60°C incubator

Procedure

1 Separate phosphoglucomutase isozymes in a slab gel by isoelectric focusing according to published methods [47].

2 Prepare a 7.6 ml overlay matrix by adding 1.6 ml of Tris buffer solution to 6 ml of distilled water. Take half of this, add 60 mg agarose, melt and allow to cool to 60°C. Adjust the size of the overlay matrix to match the size of the separating gel, keeping the overlay about 1 mm thick.①

3 To the other half, add 350 mg glucose-1-phosphate, 2.5 mg glucose-1,6-bisphosphate, 50 mg $NADP^+$, 10 mg PMS, 100 μl glucose-6-phosphate dehydrogenase (1 mg/ml) and 25 mg MTT.①

Notes

This procedure will take approximately 1 hour.

① See notes 3 and 4 of *Protocol 4* for melting agarose, and *Protocol 5d* for casting an agarose gel on to Gelbond backing. Excessive heating will denature glucose-6-phosphate dehydrogenase.

② Prepare the indicator matrix just prior to use.

③ The concentration of reagents is much higher than would be used in solution. This permits a short incubation time, which limits diffusion. This strategy is adaptable to many zymographic identifications.

4 Mix products of steps 2 and 3 together and rapidly pour on to a Gelbond backing sheet. Allow to solidify.① ②

5 Abut the indicator matrix against the separating gel and incubate for 2–4 min at 60°C. ③

Protocol 25. **Activity gels**

Reagents

Solutions from *Protocol 12*

Substrate solution – Dissolve 20 mg of fibrinogen or gelatin or casein in 2 ml of distilled water. For casein, raise the pH by adding one drop of 2 M Tris-HCl

Tris buffer – 50 mM Tris-HCl, pH 7.4

Tris-Triton solution – Tris buffer containing 2.5% (v/v) Triton X-100

Triton solution – 2.5% (v/v) Triton X-100 in distilled water

Equipment

Rocking platform

Staining dish

Procedure ① ②

1 Prepare an 8%T 2.66%C SDS slab activity gel by modifying the recipe given in *Protocol 12* as follows. To make a 12 ml gel, mix:
- 3 ml of acrylamide/bisacrylamide solution
- 2.4 ml of 5× separating gel buffer
- 5.3 ml of distilled water
- 1.2 ml of substrate solution.

2 Cast and run the gel according to *Protocol 12*. ③ ④

3 Following electrophoresis, wash the gel for 20 min in five volumes of Triton solution, then for 20 min in Tris-Triton, then for 20 min in Tris buffer (room temperature on a rocking platform). Increase incubation times for gels thicker than 0.75 mm. ⑤

Notes

This procedure will take approximately 11 hours.

① This procedure demonstrates neutral proteolytic activity in an SDS activity gel.

② To convert this to an overlay indicator matrix procedure: (i) replace the substrate solution in the separating gel recipe with distilled water; (ii) cast an overlay gel which is 1% (w/v) agarose in Tris buffer containing 1 mg/ml protein substrate *or* 7%T, 2.66%C polyacrylamide in Tris buffer containing 1 mg/ml protein substrate.

③ If possible, use an activity gel the same day it is cast. Overnight storage is not recommended.

④ Do not heat the samples prior to electrophoresis. For some enzymes, it is necessary to omit sulfhydryl agents from the SDS-sample buffer as well.

4 Incubate the gel with rocking for 1–4 h at 37°C in five volumes of Tris buffer. ⑥

5 Fix and stain according to *Protocol 17a*. ⑦

⑤ A high concentration of Triton X-100 in the washing solutions displaces the SDS from the proteins.

⑥ Many proteolytic enzymes can be demonstrated without adding cofactors. Some may require sulfhydryl agents [40].

⑦ Molecular weight markers will not be visible in these gels. To assess molecular weight of the bands and to assess the presence of other proteins in the sample, run a companion gel without co-polymerized protein. Compare the two to identify the band containing the enzyme.

VII RECORDING AND STORING GELS

Introduction

Following electrophoresis and demonstration of the proteins, a permanent record of the results is usually required. This can be achieved by preserving the gel directly or by reproducing the pattern in another medium. The most versatile means (using the costliest equipment) is to convert the electrophoresis pattern to a digital image which can then be stored *in toto*, abstracted or processed.

Methods available

Computerized methods

For slab gels or blots containing radioactive and some chemiluminescent reporters, two-dimensional imagers are commercially available (e.g. Bio-Rad and Molecular Dynamics/Eastman Kodak). In these systems, a large slab gel or blotting membrane (e.g. 20 × 25 cm) is laid upon a position-sensitive detector plate. Counts per minute are simultaneously accumulated in adjacent pixels across the entire area of the plate. A printout of c.p.m. by address or a reconstructed digital image is generated by software. For slab gels stained with soluble, colloidal or metal stains, many home-made and commercially available (e.g. Bio-Image or Image Master systems) arrangements have been described to create and analyze digital images. Some of these systems use laser scanners while others use television or CCD cam-

References

1. Tal, M., Silberstein, A. and Nusser, E. (1985) *J. Biol. Chem.* **260**, 9976–9980.
2. Rodriquez, L.V., Gersten, D.M., Ramagli, L.S. and Johnston, D.A. (1993) *Electrophoresis*, **14**, 628–637.
3. Sammons, D.W., Adams, L.D. and Nishizawa, E.E. (1981) *Electrophoresis*, **2**, 135–141.
4. Gersten, D.M., Zapolski, E.J. and Ledley, R.S. (1983) *Analyt. Biochem.* **129**, 57–59.

Protocols provided

26a. *Recording gel patterns using the Electrophoresis Duplicating System*
26b. *Photographing gels*
27. *Drying a polyacrylamide gel on to filter paper*

eras to accumulate an image of gels or autoradiography films trans-illuminated on a light box. The camera-based systems can also digitize opaque patterns (blots or gels dried on to opaque supports and photographs) with illumination from above.

Once the band or spot pattern is converted into a digital image, quantification of the pattern by computerized densitometry is straightforward; extraction of meaningful quantitative data from the result is not. The factors confounding interpretation of the results have less to do with hardware and software issues and more to do with the nature of the demonstration methods themselves. That is, the assignation of actual mass to a given protein band or spot having a measurable stain (or reporter) intensity, is the most difficult aspect of gel analysis. The reasons for this are many, but are readily illustrated by some specific examples.

(i) Stain binding by proteins depends primarily on their content of basic amino acids [1].
(ii) For silver staining, the rate reaction for grain precipitation differs widely among proteins [2].
(iii) For radioactive proteins, metabolic labeling efficiency is variable among proteins and covalent labeling is amino acid composition-dependent.

Thus, unequal amounts of different proteins can give the same intensity and vice versa. Difficulties inherent in these procedures are further compounded when additional steps such as photography or fluorography are added. Hence, while not always practical, analysis of 'wet' gels is preferable.

Unidimensional scanning

Unidimensional scanning of gels can be accomplished using less expensive means. For example, radioactive gels can be fractionated sequentially and the slices counted electronically. Gels in which the protein bands are demonstrated by staining can be scanned with dedicated scanners (e.g. Ultroscan laser densitometer) or with spectrophotometers. Gel-scanning attachments to spectrophotometers consist of a cuvette large enough to hold a gel about 12 cm long and a motorized platform which moves the gel past the photocell. The optical density is plotted by a strip-chart recorder. This arrangement was designed originally for cylindrical gels, but slab gels can also be scanned, if cut into individual lanes. Slab gels scanned in this way should be dried on to their plastic backing sheets, between transparent cellophane sheets, or on to a dialysis membrane (see *Protocol 27*). The dried strip is inserted into the cuvette and taped to its wall with transparent tape. An alternative spectrophotometric procedure is to slice the gel sequentially, elute the Coomassie Blue from the slices, and read the optical density of the eluate [1].

Photocopying, photographing and duplicating (see *Protocol 26*)

The least expensive and most commonly used means of transferring the gel pattern to another medium are to photocopy or photograph the gel. Both dried and 'wet' gels can be photocopied or photographed. Of the two techniques, photocopying, using an ordinary photocopier, is the most convenient and least expensive way to obtain a permanent record of the gel pattern. The record, however, will not be of sufficient quality for publication, or for subsequent conversion to a digital image.

To produce pictures of electrophoresis gels, two alternatives are available. The first uses a specialized film for reproducing silver-stained patterns and a specialized paper for Coomassie Blue-stained patterns. The Electrophoresis Duplicating System (Eastman Kodak) is a direct duplicating system without a camera, specifically dedicated to gel work (see *Protocol 26a*). The second alternative uses a camera and ordinary photographic supplies (see *Protocol 26b*). Gel photography for maintaining a permanent record of the electrophoresis, for publication of the results or for subsequent computer analysis is a straightforward procedure. An ordinary camera, a copying stand to hold the camera and a photographic light box are required. The gels can be photographed using conventional or 'instant' (e.g. Polaroid) film. Since the procedures use readily available film, it is less expensive to have the film developed commercially than to assemble one's own developing facility. The procedures given are standard and will serve the majority of applications.

Drying (see *Protocol 27*)
For maintaining a permanent record, preserving polyacrylamide and agarose slab gels stained with soluble or colloidal stains is common practice. Preserving metal-stained gels is less common because some metal stains (e.g. some silver formulations) fade with time [3]. Preserving cylindrical gels is rarely done. Agarose slab gels are easily dried by pressing. To dry an agarose slab gel, cover the gel with Whatman 3MM filter paper and a pad (about three times as thick as the gel) of absorbent tissue. Place a glass plate and a weight on top and blot the liquid out of the gel. When no more liquid can be removed by blotting, allow the gel to air dry. A completely dry gel will be paper thin and have the consistency of delicate cellophane. Store the dried gel at room temperature in a glassine envelope indefinitely.

Polyacrylamide gels are more difficult to dry because they shrink in all dimensions when water is removed. Accordingly, a gel of 10%T, will, if allowed to air dry, reduce to about 25% of its original volume. In addition, if air drying is too rapid or uneven, the gel will crack. Thus, gels thinner than about 3–4 mm are not easily air dried. To air dry, first soak the gel in fixing solution (*Protocol 17a*) containing 5% (v/v) glycerol. Lay the gel on a glass plate and turn it every 12 h for about a week. The result will be a hard plastic tile with a sticky surface from the glycerol. Wrap the gel in plastic wrap and store at room temperature. To maintain the original length and width dimensions, polyacrylamide slab gels are most frequently dried paper thin on to a support matrix. The matrix can be the silanized glass electrophoresis plate, Gelbond plastic backing sheet (see *Protocol 4*), a dialysis membrane, cellophane or, most commonly, acid-washed filter paper. Gels covalently bonded to silanized glass or plastic can be allowed to air dry (rehydratable ultrathin gels can be oven-dried at 50–60°C) while those dried on to the other matrices require special arrangements. In common practice, drying is usually accelerated by heating the gel.

There are three basic types of gel drier. The first dries the gel between two sheets of cellophane by blowing hot air across the surfaces of the cellophane. The second is a vacuum–hot plate combination which heats from one surface and dries the gel on to filter paper or dialysis membrane. The vacuum–hot plate combination is the most common instrument. The third is a vacuum–microwave combination which heats the gel from within and dries the gel rapidly on to filter paper or dialysis membrane. The first and second types of instrument are commercially available and work well. The microwave driers which are commercially available are unsatisfactory. Microwave driers which dry a 0.75 mm gel in about 3.5 min can be easily constructed according to the instructions published elsewhere [4].

Protocol 26a. **Recording gel patterns using the Electrophoresis Duplicating System**

Reagents

Dektol developer (Eastman Kodak)
Rapid Fix (Eastman Kodak)

Equipment

Darkroom with yellow or red safelights
Electrophoresis duplicating film (Eastman Kodak)
Electrophoresis duplicating paper (Eastman Kodak)
Fluorescent lamp
Glass plate
Incandescent lamp

Procedure

Silver stained gels

1 In a darkroom with a red or yellow safelight, place the film on a clean, dry glass plate. Dirt and smudges will become part of the image, as will fingerprints on the film.

2 Cover the film with the acetate sheet provided with it, then place the wet gel on top of the cellulose acetate sheet. Remove any air bubbles from between the gel and acetate.

3 Position a fluorescent lamp, equipped with two 15 W cool-white bulbs

Notes

This procedure will take approximately 1 hour.

① Exposure time varies with light intensity and contrast between stained proteins and background. For silver, grayness of the gel background is a consequence of washing following silver staining and of differences in silver stain formulations. For soluble stains, contrast is a function of destaining. Determine the correct time by exposing trial strips of film for various intervals. This is necessarily repeated when the lamp is moved and as fluorescent bulbs age.

(this can be an ordinary desk lamp), 20–26 cm above the gel.

4 Turn on the fluorescent lamp (with a hand-held switch or at the wall socket to avoid shaking the lamp and table) and expose the film for 2–3 min.①

5 Remove the film and develop according to the manufacturer's instructions.②

6 Hang the film up to air dry.

Coomassie Blue-stained gels

1 In a darkroom with a yellow safelight, place a glass plate on a level table and position a 15 W incandescent bulb 90 cm above it.

2 Place a sheet of photographic paper on the glass plate and cover it with the amber filter provided with the photographic paper. Then place the wet gel on top of the amber filter sheet. Remove any air bubbles from between the gel and filter.

3 Turn on the incandescent light (with a hand-held switch or at the wall socket to avoid shaking the lamp and table) and expose the paper/gel sandwich for 5–10 sec.①

4 Remove the paper and develop according to the manufacturer's instructions.③

5 Hang the paper up to air dry.

② Developing program is: 1 min at 70°C in Kodak Dektol developer (diluted 1:1 in water), 10–20 sec rinse in running water, 1–2 min in Kodak Rapid Fix at room temperature, 10–20 min rinse in running water.

③ Developing program is: 1 min at 70°C in Kodak Dektol developer (diluted 1:1 in water), 30 sec rinse in running water, 1–2 min in Kodak Rapid Fix at room temperature, 10 min rinse in water.

Protocol 26b. **Photographing gels**

Equipment

Camera with cable release (35 mm camera with a 50–55 mm macro lens or 4 × 5 camera with a 135 or 150 mm lens)

Film ①

Light box ②

Photographic copy stand or laboratory ring stand

Procedure ③

1 Place the gel on a light box and remove any trapped air bubbles. Place a transparent ruler alongside the gel in the direction of electrophoresis. The picture will be a different size to the gel; photographing the ruler will allow measurement of migration distance directly from the print or negative.④

2 Attach the appropriate filter to the camera – yellow, dark yellow or orange for Coomassie-stained gels (Kodak no. 16 or equivalent), green for silver-stained gels (Kodak no. 58 or equivalent).

3 Mount the camera on the stand and adjust the distance from the light box such that the band or spot pattern occupies the entire field. Focus on the bands, not on the ruler.⑤

4 Set the aperture opening and shutter speed then stand back from the

Notes

This procedure will take approximately 1 hour.

① Black and white film is recommended for most gel photography since it will give the widest range of gray levels. Use Kodak Ektapan 100 or Tri X 320, or their equivalents from other manufacturers. When sharpness of image is more important than number of gray levels, use high contrast process film no. 4514. If 'instant' film is desired, use Polaroid PRO 100 (4″ × 5″ format). When photographing gels stained with single colors other than Coomassie Blue and silver-gray, use black and white film and try using an orange/yellow filter. When proteins have different colors (e.g. colored silver stains) use Kodak 64 color film without a filter.

② Most photographic light boxes use fluorescent lamps but incandescent lamps fitted with rheostats allow easy adjustment of light intensity.

table to allow vibration to stop. Snap the picture using a cable release to avoid further vibration. See note 6 for settings. ⑥

5 Develop the film according to the manufacturer's instructions.

③ To photograph a silver-stained or Coomassie Blue-stained 'wet' gel, or one dried on to a transparent backing: see note 7 for cylindrical gels, note 8 for gels dried on to opaque supports and note 9 when photographing for computer analysis .

④ The gel will tend to dry out and curl if left on the light box too long. Keep it moistened with destaining solution.

⑤ The most accurate image is the one which occupies the most film. Adjust the size of the ultimate picture by adjusting the printing rather than the filming.

⑥ Exposure conditions vary with light intensity and contrast between the stained bands or spots and clarity of the gel background. Determine the correct settings by 'bracketing'. For black and white film, start with the camera set at f16, and 1/15 sec. Go up and down at least one stop for each.

⑦ Place the cylindrical gel in a test tube, fill it completely with destaining solution and stopper it. Trans-illuminate by laying the test tube down on the light box.

⑧ Flatten the gel by covering it with a glass plate and illuminate it from above. If illumination from above causes glare, cover the gel with frosted rather than clear glass.

⑨ (i) The camera lens must be accurately parallel to the gel surface. Use a carpenter's level. (ii) The amount of light coming through the gel at all points must be even. Measure the intensity at the gel surface with a light meter. If the intensity varies by more than a few per cent, use a light box with more closely spaced tubes *or* place a frosted glass plate between the box and gel *or* place the gel on a glass plate and raise the plate away from the surface of the box.(iii) Ensure that there is no glare.

Protocol 27. **Drying a polyacrylamide gel on to filter paper**

Equipment

Acid-washed filter paper (Whatman 3MM or equivalent)
Gel dryer (vacuum/hot plate style, e.g. BioRad model 543 or equivalent) ①
Vacuum pump and trap

Procedure

1 Place the porous polyethylene sheet at the bottom of the dryer.

2 Cut a piece of acid-washed filter paper (Whatman 3MM or equivalent) to a size about 4 cm larger in both dimensions than the gel, saturate it with water and place it on the porous polyethylene. ②

3 Place the gel on top of the wet filter paper and press out any air bubbles from between the gel and paper.

4 Cover the gel with the cover sheet then with the transparent silicone rubber vacuum sealing sheet.

5 Connect the vacuum line and draw the vacuum. When you can no longer see liquid being drawn into the vacuum trap, close the cover and set the

Notes

This procedure will take approximately 1–3 hours, depending on thickness (see ④).

① This dryer heats the gel from the top and pulls the vacuum from the bottom. For those which heat from the bottom and draw the vacuum from the top, reverse the order of the 'stack'.

② To prevent clogging, soak the porous polyethylene in water after use. When drying gels for autoradiography (not fluorography – see *Protocol 20*), use two additional sheets below the gel and two between the gel and the cover sheet to ensure that the dryer does not become contaminated.

③ Protect the pump from water and steam damage with a cold trap (e.g. an ordinary side-arm flask sitting in an ice bucket).

temperature for 80°C (60°C for gels impregnated with PPO for fluorography) and turn on the heater.③ ④ [1]

6 When drying is complete, turn off the heater. The gel will have the consistency of soft plastic. In order to avoid curling of the dried gel, either leave the vacuum on until the gel has fully cooled, or put the dried gel under a heavy weight until cool. The dried gel can be stored indefinitely at room temperature.

④ Do not break the vacuum until the gel is fully dried or else it will crack irreparably. Drying time depends on gel thickness, temperature and strength of the vacuum. A 0.75 mm thick gel dried at 80°C with a vacuum of 20 lb dries in about 45 min. Open the cover with the vacuum still engaged. The gel is dry when the height of the 'stack' no longer changes.

Pause point

[1] At 80°C, it is virtually impossible to overdry the gel.

APPENDIX A: SOLUTION RECIPES

Acrylamide/AcrylAide stock: 30 g acrylamide ⚠, 40 ml of 2% (w/v) AcrylAide solution (FMC Bioproducts) made up to 100 ml with distilled water. Store refrigerated.

Acrylamide/bisacrylamide stock for all gels except *Protocols 11* and *13a*: 30 g acrylamide ⚠, 0.8 g bisacrylamide ⚠ made up to 100 ml with distilled water. Store at room temperature.

Acrylamide/bisacrylamide stock for *Protocol 11*: 29.1 g acrylamide ⚠, 0.9 g bisacrylamide ⚠ made to 100 ml with water. Store at room temperature.

Acrylamide/ bisacrylamide for the first dimension of *Protocol 13a*: 30 g recrystallized acrylamide ⚠, 1.8 g bisacrylamide ⚠ made up to 100 ml with distilled water. Store at room temperature.

5% w/v Ammonium persulfate: Make up with distilled water and store refrigerated, for no longer than 1 week.

Ampholyte solution for *Protocol 10*: Narrow range (e.g. pH 5–7). Dilute the Ampholine solution from the manufacturer to 4% (v/v) 5–7 Ampholine in 10% (v/v) glycerol. Store in small aliquots at −70°C.

Ampholyte solution for *Protocol 10*: Wide range (e.g. pH 3–10) Mix 5–7 Ampholine and 3–10 Ampholine in a ratio of 4:1 (v/v). Dilute the mixture to a final concentration of 4% (v/v) in 10% glycerol. Store in small aliquots at −70°C.

Barbital buffer for immunoelectrophoresis (0.05 M): 1.84 g of barbital ⚠ and 10.3 g of sodium barbital ⚠ made up to 1 liter with distilled water. Adjust to pH 8.6. Store at 4°C for as long as the pH is maintained.

Binding buffer solution with t-RNA for *Protocol 21c*: 50 mM NaCl, 1 mM EDTA, 0.02% (w/v) bovine serum albumin, 0.02% (w/v) Ficoll 400, 0.02% (w/v) polyvinylpyrrolidone 40, 20 μg/ml *E. coli* t-RNA (Sigma). Store frozen in individual aliquots.

CHAPS sample buffer for isoelectric focusing: 2% CHAPS (full name is 3-[3-(chloramidopropyl) diethylammonio]-1-propane sulfonate), 0.5% (w/v) Ampholytes, pH 5–7.

Elution buffer for immunoprecipitation: 0.15 M NaCl, 0.1 M CH_3COOH, pH 2.9. Store at room temperature.

Equilibrating solution: 36 g recrystallized urea, 30 ml glycerol, 2 g SDS, 10 ml 0.5 M Tris-HCl, pH 6.8, 5 mg Bromophenol Blue, made up to 100 ml in distilled water. Store in individual aliquots at −70°C.

First-dimensional overlay solution for *Protocol 13a*: 2 ml 9 M urea, 40 μl Ampholine 5–7, 5 μl Ampholine 3–10. Store frozen in small aliquots.

Rehydrating solution for *Protocol 13a*: 0.5% (w/v) carrier ampholytes of the same range of the Immobiline gradient, 0.5% (v/v) NP-40, 0.2% (w/v) dithiothreitol made up in deionized 8 M urea. Store in individual aliquots at −70°C.

Running (tank) buffer for SDS gels: 3.027 g Tris-base, 14.41 g glycine, 1.0 g SDS. Adjust to pH 8.3 and make up to 1 liter with distilled water. Store at room temperature for as long as the pH is maintained.

Running (tank) buffer for Tris-glycine gels: 6.0 g Tris-base, 28.8 g glycine made up to to 1 l with distilled water. Adjust to pH 8.3. Store at room temperature for as long as the pH is maintained.

5× Sample buffer solution for Tris-glycine gels: Mix 15.5 ml of 1 M Tris-HCl (pH 6.8) with 2.5 ml of 1% (w/v) Bromophenol Blue, 7 ml of distilled water and 25 ml of glycerol. Store refrigerated for as long as the pH is maintained.

Saturated solution of Oil Red O for *Protocol 17e*: Make up in 60% (v/v) ethanol and store at 37°C for no longer than a few weeks.

SDS-CHES solution: 100 mg CHES [Calbiochem (full name is cyclohexylaminoethane sulfonic acid)], in 7 ml distilled water. Adjust to pH 9.5 with 0.5 M NaOH then add 200 mg SDS and 0.5 ml of 2-mercaptoethanol or 100 mg of dithiothreitol. Can be stored at room temperature for as long as the pH is maintained.

SDS sample buffer: 0.98 g Tris-HCl, 2.0 g SDS, 7.5 ml glycerol, 5 ml 2-mercaptoethanol. Adjust pH to 6.8 and make up to 100 ml with distilled water. Store at room temperature for as long as the pH is maintained.

2× SDS sample buffer: 0.98 g Tris-HCl, 2.0 g SDS, 7.5 ml glycerol, 5 ml 2-mercaptoethanol. Adjust to pH 6.8 and make up to 50 ml with distilled water. SDS sample buffer can be stored at room temperature for as long as the pH is maintained.

5× Separating gel buffer for SDS gels: 226.9 g Tris base, 1.25 ml TEMED ⚠, 5.0 g SDS. Adjust pH to 8.8 and make up to 1 liter with distilled water. Store at room temperature for as long as the pH is maintained.

5× Separating gel buffer for Tris-glycine gels: 22.7 g Tris base, 140 μl TEMED ⚠ in 100 ml of 0.48 M HCl (adjust to pH 8.9 if necessary). Store at room temperature for as long as the pH is maintained.

5× Stacking gel buffer for SDS gels: 9.85 g Tris-HCl, 0.125 ml TEMED ⚠, 0.5 g SDS. Adjust to pH 6.8 and make up to 100 ml with distilled water. Store at room temperature for as long as the pH is maintained.

5× Stacking gel buffer for Tris-glycine gels: 7.48 g Tris-base, 280 μl TEMED ⚠ in 100 ml of 0.48 M HCl (adjust to pH 6.7 if necessary). Store at room temperature for as long as the pH is maintained.

Sudan Black B solution for *Protocol 17e*: 100 mg in 200 ml of 60% (v/v) ethanol. Store at room temperature for no longer than a few weeks.

Tris-buffered saline (TBS): 20 mM Tris-HCl, 0.145 M NaCl, 1 mM EDTA, 10 μg/ml aprotinin, pH 7.2. Store at room temperature and add aprotinin immediately before use.

8 M Urea: Dissolve 480.5 g of analytical grade urea in distilled water and make up to 1 liter. Pass over a bed of Dowex mixed-bed ion-exchange resin (MBL). Store at room temperature.

Urea–NP-40 solution: 5.4 g recrystallized urea, 0.4 ml NP-40, 0.2 ml Ampholytes pH 3–10, 0.5 ml 2 mercaptoethanol. Make up to 10 ml with distilled water. Can be stored refrigerated for 3 days or indefinitely at −70°C.

All fixing, staining and destaining solutions can be stored for several weeks at room temperature.

APPENDIX B: MARKER PROTEINS FOR MOLECULAR WEIGHTS AND pI

Preparation of radioactive inks and prestained markers for fluorography experiments

Markers indicating molecular weight, pI, lane orientation, etc., will only appear on the film record if they are themselves radioactive. For convenience, mark the stained, dried gel with a radioactive ink, in the position of the markers. To prepare the ink, mix India ink (Higgins) with the radioisotope used in the samples at 10^6 d.p.m./ml. Then cover the marker lane of the dried gel with a strip of tape (Scotch Magic Tape or equivalent) and apply the radioactive ink to the tape. Prestained molecular weight markers are useful when staining of the entire gel is not required. Prestain using Drimarene Brilliant Blue K-BL (also called Reactive Blue 114) according to ref. 29 of Chapter V.

Molecular weight marker kits

Pharmacia, 14–94 kDa:

α-Lactalbumin	14.4
Trypsin inhibitor	20.1
Carbonic anhydrase	30.0
Ovalbumin	43.0
Serum albumin	67.0
Phosphorylase B	94.0

Pharmacia native, 67–669 kDa:

Serum albumin	67.0
Lactate dehydrogenase	140.0
Catalase	232.0
Ferritin	440.0
Thyroglobulin	669.0

Pharmacia reduced, 53–212 kDa:

Glutamate dehydrogenase	53.0
Transferrin	76.0
β-Galactosidase	116.0
α-2-Macroglobulin	170.0
Myosin	212.0

Pharmacia myoglobin peptides, 2.512–17 kDa:

2.512
6.214
8.159
10.7

	14.404
	16.949

BioRad, 19–107 kDa:

Lysozyme	19.0
Trypsin inhibitor	27.2
Carbonic anhydrase	36.8
Ovalbumin	52.0
Bovine serum albumin	76.0
Phosphorylase B	107.0

BioRad, 47–205 kDa:

Ovalbumin	47.0
Bovine serum albumin	85.0
β-Galactosidase	118.0
Myosin	205.0

BioRad prestained markers – same color :
mixture of myosin, β-galactosidase, bovine serum albumin, ovalbumin, carbonic anhydrase, trypsin inhibitor, lysozyme, aprotinin.

BioRad prestained markers – different colors :
polypeptide mixture of carbonic anhydrase, trypsin

α-2-Macroglobulin	
reduced	170.0
nonreduced	340.0

Amersham ^{14}C methylated radioactive markers
High molecular weight, 14–200 kDa: lysozyme, carbonic anhydrase, ovalbumin, bovine serum albumin, phosphorylase B, myosin.

Low molecular weight, 3.4–30 kDa: insulin A chain, insulin B chain, aprotinin, cytochrome *c*, trypsin inhibitor, carbonic anhydrase.

Amersham radioactive step wedges for fluorography
^{3}H standards : range of 3–110 nCi/mg

^{14}C standards: low range = 0.1–100 nCi/g, high range = 31–883 nCi/mg

^{125}I standards: range of 1.25–640 nCi/mg

Isoelectric point marker kits

Pharmacia broad pI:

Amyloglucosidase	3.50
Methyl Red dye	3.75

inhibitor, lysozyme, aprotinin, insulin.

BioRad broad range mixture:
Myosin, β-galactosidase, bovine serum albumin, carbonic anhydrase, trypsin inhibitor, lysozyme, aprotinin.

BioRad biotinylated markers for detection by biotin-avidin systems :
Same molecular weight ranges as unlabeled markers.

Boehringer Mannheim, 3.4–21 kDa:

Insulin B chain	3.4
Aprotinin	6.5
Cytochrome *c*	12.5
Trypsin inhibitor	21.0

Boehringer Mannheim, 14.3–340 kDa:

Lysozyme	14.3
Trypsin inhibitor	20.1
Triose phosphate isomerase	26.6
Aldolase	39.2
Glutamate dehydrogenase	55.6
Fructose-6-phosphate kinase	85.2
β-Galactosidase	116.0

Trypsin inhibitor	4.55
β-Lactoglobulin A	5.20
Carbonic anhydrase (cow)	5.85
Carbonic anhydrase (human)	6.55
Myoglobin acidic band	6.85
Myoglobin basic band	7.35
Lentil lectin acidic	8.15
Lentil lectin middle	8.45
Lentil lectin basic	8.65
Trypsinogen	9.30

Pharmacia low pI:

Pepsinogen	2.80
Amyloglucosidase	3.50
Methyl Red dye	3.75
Glucose oxidase	4.15
Trypsin inhibitor	4.55
β-Lactoglobulin A	5.20
Carbonic anhydrase (cow)	5.85
Carbonic anhydrase (human)	6.55

Pharmacia high pI:

β-Lactoglobulin A	5.20
Carbonic anhydrase (cow)	5.85
Carbonic anhydrase (human)	6.55

Myoglobin acidic band	6.85
Myoglobin basic band	7.35
Lentil lectin acidic	8.15
Lentil lectin middle	8.45
Lentil lectin basic	8.65
Trypsinogen	9.30
Cytochrome *c*	10.25
BioRad 4.4–9.6:	
Phycocyanin acidic band	4.45
Phycocyanin middle band	4.65
Phycocyanin basic	4.75
β-Lactoglobulin	5.10
Carbonic anhydrase (cow)	6.00
Carbonic anhydrase (human)	6.50
Myoglobin	7.00
Hemoglobin A	7.10
Hemoglobin C	7.50
Lentil lectin acidic	7.8
Lentil lectin middle	8.0
Lentil lectin basic	8.20
Cytochrome *c*	9.6

APPENDIX C: SUPPLIERS

Instruments

Accurate Chemical and Scientific Corp., 300 Shames Drive, Westbury, NY 11590, USA.
Tel (800) 645 6264, (516) 333 2221. Fax (516) 997 4948.

Ambis, 3939 Ruffin Road, San Diego, CA 92123, USA.
Tel (619) 571 0113. Fax (619) 571 5940.

ATTO Instruments, 1500 Research Blvd, Rockville, MD 20850, USA.
Hongo 7-2-3, Bunkyo-ku, Tokyo 113, Japan.
Tel 3 3814 4861. Fax 3 5684 6646.

Bio-Rad Laboratories, 2000 Alfred Nobel Drive, Hercules, CA 94547, USA.
Tel (510) 741 1000. Fax (510) 741 1060.
Bio-Rad House, Maylands Avenue, Hemel Hempstead, Herts HP2 7TD, UK.
Tel (0800) 181134. Fax (01442) 259118.

British Biotech Pharmaceuticals Ltd, Watlington Road, Oxford OX4 5LY, UK.
Tel (01865) 748747. Fax (01865) 781115.

Dupont-NEN, 549 Albany Street, Boston, MA 02118, USA.
Tel (617) 482 9595. Fax (617) 542 8463.
Wedgwood Way, Stevenage, Herts SG1 4QN, UK.
Tel (01438) 734027. Fax (01438) 734049.

E-C Apparatus, 3831 Tyrone Blvd North, St Petersburg, FL 33709, USA.

Electrobiotransfer, PO Box 9847, San Jose, CA 95157-0847, USA.
Tel (415) 327 4067. Fax (415) 327 8564.

Fisher Scientific, 711 Forbes Ave, Pittsburgh, PA 15219, USA.
Bishop Meadow Road, Loughborough, Leics LE11 5RG, UK.
Tel (01509) 231166. Fax (01509) 231893.

Gibco-BRL, PO Box 6009, Gaithersburg, MD 20877, USA.
PO Box 35, Trident House, Renfrew Road, Paisley PA3 4EF, UK.
Tel (0141) 814 6100. Fax (0141) 814 6258.

ISCO, PO Box 5347, Lincoln, NE 68505, USA.
Tel (402) 464 0231. Fax (402) 464 4543.
c/o Jones Chromatography, New Road, Hengoed, Mid-Glamorgan, UK.
Tel (01443) 816991. Fax (01443) 816816.

Integrated Separation Systems, 1 Westinghouse Plaza, Hyde Park, MA 02136, USA.

Labcanco, Prospect Avenue, Kansas City, MO, USA.
Tel (816) 333 8811. Fax (816) 363 0130.

Molecular Dynamics, 928 East Arques Avenue, Sunnyvale, CA 94086, USA.
Tel (800) 333 5703. Fax (408) 773 8348.
4 Chaucer Business Park, Sevenoaks, Kent TN15 6PL, UK.

Pharmacia Biotechnology, 23 Grosvenor Road, St Albans, Herts AL1 3AW, UK.
Tel (01727) 81400. Fax (01727) 814020.
800 Centennial Avenue, PO Box 1327, Piscataway, NJ 08855-1327, USA.
Tel (201) 457 80000. Fax (201) 457 0557.

Scientific Imaging Systems, 36 Clifton Road, Cambridge CB1 4ZR, UK.
Tel (01223) 242813. Fax (01223) 243036.

Shandon, 93 Chadwick Road, Runcorn, Cheshire WA7 1PR, UK.
Tel (01928) 566611. Fax (01928) 565845.

Consumables

Accurate Chemical and Scientific, 300 Shames Drive, Westbury, NY 11590, USA.
Tel (516) 333 2221. Fax (516) 997 4948.

Calbiochem, 10394 Pacific Center Court, San Diego, CA 92121, USA.
Tel (800) 854 3417. Fax (619) 453 3552.
Boulevard Industrial Park, Padge Road, Beeston, Nottingham NG9 2JR.
Tel (0800) 622935, (0115) 9430840. Fax (0115) 9430951.

Caltag, 1849 Bayshore Blvd, Suite 200, Burlingame, CA 94010, USA.
Tel (415) 652 0468. Fax (415) 652 9030.

Cambridge Electrophoresis, 84 High Street, Cherry Hinton, Cambridge CB1 4HZ, UK.
Tel (01223) 249431. Fax (01223) 411803.

Chemicon International, 28835 Single Oak Drive, Temecula, CA 92590, USA.
Tel (909) 676 8080. Fax (909) 676 9209.
2 Bonnersfield Lane, Harrow HA1 2JR, UK.
Tel (0181) 863 0415. Fax (0181) 863 0416.

Dako, 6392 Via Real, Carpentina, CA 93013, USA.
Tel (804) 466 6644. Fax (805) 566 688.
16 Manor Courtyard, Hughenden Avenue, High Wycombe, Bucks HP13 5RE, UK.
Tel (01494) 452016. Fax (01494) 441846.

Amersham, 2636 South Clearbrook Drive, Arlington Heights, IL 60005, USA.
Tel (708) 593 6300.
Lincoln Pl, Green End, Aylesbury, Bucks HP20 2TP, UK.
Tel (01494) 544000. Fax (01494) 542266.

A-T Biochem, 30 Spring Mill Drive, Malvern, PA 19355, USA.
Tel (610) 889 9300. Fax (610) 889 9304.

Atlanta Biologicals, 520 Pinnacle Court, Norcross, GA 30071, USA.
Tel (404) 446 3336. Fax (404) 446 1404.

Bio-Rad Laboratories, 2000 Alfred Nobel Drive, Hercules, CA 94547, USA.
Bio-Rad House, Maylands Avenue, Hemel Hempstead, Herts HP2 7TD, UK.
Tel (01442) 232522. Fax (01442) 259118.

Boehringer Mannheim, 9115 Hague Road, Indianapolis, IN 46250, USA.
Tel (800) 262 1640. Fax (317) 578 7339.
Bell Lane, Lewes, East Sussex BN7 1LG, UK.
Tel (01273) 480444. Fax (10273) 480266.

Bradsure Biologicals, 67a Brook Street, Shepshed, Loughborough, Leics LE12 9RF, UK.
Tel (01509) 650665. Fax (01509) 650544.

Dupont-NEN, 549 Albany Street, Boston, MA 02118, USA.
Tel (617) 482 9595. Fax (617) 542 8463.
Wedgwood Way, Stevenage, Herts SG1 4QN, UK.
Tel (01438) 734027. Fax (01438) 743049.

E-C Apparatus, 100 Colin Drive, Holbrook, NY 11741-4306, USA.
Tel (516) 244 2929. Fax (516) 244 0606.

Eastman Kodak – Scientific Imaging Systems, Building 642, 343 State Street, Rochester, NY 14652, USA.

Fisher Scientific, 711 Forbes Avenue, Pittsburgh, PA 15219, USA.
New Enterprise House, St Helen's Street, Derby DE1 3GY, UK.

Fluka Chemical, PO Box 14508, St Louis, MO 63178, USA.

FMC Bioproducts, 191 Thomaston Street, Rockland, ME 04841, USA.
Tel (207) 594 3400. Fax (207) 594 3491.
c/o Flowgen Instruments, Broad Oak Road, Sittingbourne, Kent ME9 8AQ, UK.

Gibco-BRL, PO Box 6009, Gaithersburg, MD 20877, USA.
PO Box 35, Trident House, Renfrew Road, Paisley PA3 4EF, UK.
Tel (0141) 814 6100. Fax (0141) 814 6258.

ICN Biomedicals 3300 Hyland Avenue, Costa Mesa, CA 92626, USA.
Tel (800) 854 0530. Fax (800) 854 0530.
Thame Business Park Centre, Wenman Road, Thame, Oxon OX9 3XA, UK.
Tel (0800) 282474/(01844) 213366. Fax (0800) 614735/(01844) 213399.

Integrated Separation Systems, 21 Strathmore Road, Natick, MA 01760, USA.
Tel (508) 655 1500. Fax (508) 655 8501.

International Biotechnologies, PO Box 9558, New Haven, CT 06535, USA.
36 Clifton Road, Cambridge CB1 4ZR, UK.
Tel (01223) 242813. Fax (01223) 243036.

Mallinckrodt Baker Inc., 222 Red School Lane, Phillipsburgh, NJ 08865, USA.
Tel (908) 859 2151. Fax (908) 859 9318.

Millipore Corp., 80 Ashby Road, Bedford, MA 01730, USA.
The Boulevard, Ascot Road, Croxley Green, Watford WD1 8YW, UK.
Tel (01923) 816375. Fax (01923) 818297.

Polysciences, Paul Valley Industrial Park. Warrington, PA 18976, USA.

Promega , 2800 Woods Hollow Road, Madison, WI 53711, USA.
Tel (608) 274 4330. Fax (608) 277 2516.
Delta House, Chilworth Research Centre, Southampton SO16 7NS, UK.
Tel (01703) 760225, (0800) 378994. Fax (01703) 767014, (0800) 181037.

Serva, c/o Crescent Chemical, 1324 Motor Parkway, Hauppage, NY 11788, USA.

Schleicher and Schuell, 10 Optical Avenue, PO Box 2012, Keene, NH 03431, USA.
Tel (603) 352 3810. Fax (603) 357 3627.

Sigma-Aldrich Chemical, PO Box 14508, St Louis, MO 63178, USA.
Tel (314) 771 5765. Fax (314) 771 5757.
Fancy Road, Poole, Dorset BH17 7NH, UK.
Tel (0800) 373731. Fax (01202) 715460.

Upstate Biotechnology, 199 Saranac Avenue, Lake Placid, NY 12946, USA.
UK phone (0800) 894868.

Tel (508) 927 5054. Fax (508) 921 1350.
67 Knowl Piece, Wilbury Way, Hitchin, Herts SG4 0TY, UK.
Tel (01462) 420616. Fax (01462) 421057.

Pharmacia, 800 Centennial Avenue, Piscataway, NJ 08855, USA.
Tel (201) 457 8000. Fax (201) 457 0557.
23 Grosvenor Road, St Albans, Herts AL1 3AW, UK.
Tel (01727) 814000. Fax (01727) 814020.

Pierce, 3747 North Meridian Road, PO Box 117, Rockford, IL 61105, USA.
Tel (800) 874 3723/(815) 968 0747. Fax (815) 968 7316.
Pierce Europe, PO Box 1512, 3260, Oud Beijerland, The Netherlands.

Amersham Place, Little Chalfont, Bucks HP7 9NA, UK.

Whatman, 9 Birdwell Place Clifton, NJ 07014, USA.
Whatman House, St Leonard's Road, 20/20 Maidstone, Kent ME16 0LS, UK.
Tel (01622) 676670. Fax (01622) 677011.

Zymed Laboratories, 458 Carleton Court South, San Francisco, CA 94080, USA.
c/o Cambridge Bioscience, 25 Signet Court, Stourbridge Common Business Centre, Swann's Road, Cambridge CB5 8LA, UK.

INDEX